# CENTRIFUGAL
## and
# ROTARY PUMPS
## Fundamentals With Applications

## LEV NELIK

**CRC Press**
**Boca Raton   London   New York   Washington, D.C.**

Contact Editor:         Cindy Renee Carelli
Project Editor:          Sara Rose Seltzer
Marketing Managesr:   Barbara Glunn, Jane Stark,
                            Jane Lewis, Arline Massey
Cover design:           Dawn Boyd

**Library of Congress Cataloging-in-Publication Data**

Nelik, Lev.
    Centrifugal and rotary pumps : fundamentals with applications /
Lev Nelik.
       p.  cm.
    Includes bibliographical references and index.
    ISBN 0-8493-0701-5 (alk. paper)
    1. Centrifugal pumps.  2. Rotary pumps.  I. Title.
TJ919.N34 1999
621.6′7—dc21
                                                        98-49382
                                                            CIP

No claim to original U.S. Government works
International Standard Book Number 0-8493-0701-5
Library of Congress Card Number 98-49382
Printed in the United States of America  1  2  3  4  5  6  7  8  9  0
Printed on acid-free paper

# Preface

My motivation in writing this book was to relate fundamental principles of the operation of kinetic and positive displacement pumps, with direct relation to application specifics and user needs. In today's reality, pump users demand simpler, easier-to-read, and more practical material on pumps. New, young engineers who enter the workforce are faced with immediate practical challenges presented to them by the plants' environments: to solve pumping problems and improve equipment reliability and availability — in the most cost-effective manner. To meet these challenges, plant personnel must first understand the *fundamentals* of pump operations, and then apply this knowledge to solve their immediate short-term, and long-term, problems. Pumps are the most widely used type of machinery throughout the world, yet, unfortunately, they are covered very little, or not at all, at the college level, leaving engineering graduates unprepared to deal with — not to mention troubleshoot — this equipment. The variety of pump types also adds to the confusion of an engineer entering the workforce: Which pump type, among many, to choose for a given application? Available books on pumps are good but do not reflect the rapid changes taking place at the plants — tougher applications, new corrosive chemicals, and resistance to the abrasives, which because of cost pressures are no longer adequately removed from the streams before they enter a pump's suction, etc. In recent years, heightened attention to a safe workplace environment, and plants' demand for better equipment reliability have necessitated improvements in mean time between failures (MTBF), as well as a better understanding of pump fundamentals and differences — real or perceived. In addition, existing books often contain complicated mathematics with long derivations that typically make them better suited for academic researchers, not practicing engineers, operators, or maintenance personnel looking for practical advice and a real solution for their immediate needs. The emphasis of this book, therefore, is on *simplicity* — to make it useful, easy, and interesting to read for a broad audience.

For new engineers, mechanics, operators, and plant management, this book will provide a clear and simple understanding of pump types, as defined by the Hydraulic Institute (HI). For more experienced users, it will provide a timely update on the recent trends and developments, including actual field troubleshooting cases where the causes for each particular problem are traced back to pump fundamentals in a clear and methodical fashion. The pump types covered include: centrifugal, gear, lobe, vane, screw, diaphragm, progressing cavity, and other miscellaneous types.

The variation in types of pumps is presented in terms of hydraulic design and performance, principles of operation, design similarities and differences, and historical trends and technological changes. After covering fundamentals, the focus shifts to real field cases, in terms of applications, pumpage, system interaction, reliability and failure analysis, as well as practical solutions for improvements. Upon completion of the book, readers should be able to immediately implement the techniques covered in the book to their needs, as well as share what they have learned with colleagues in the field.

Existing material on pumps and pumping equipment covers predominately *centrifugal* pumps. Centrifugal pumps have dominated the overall pumping population in the past, but this situation has been changing in the last 10 to 15 years. New chemicals, industrial processes, and technologies have introduced processes and products with viscosities in ranges significantly beyond the capabilities of centrifugal pumps. Many users still attempt to apply centrifugal pumps to such unsuited applications, unaware of new available pump types and improvements in *rotary* pump designs. Furthermore, there is very little published material on gear pump designs — the effects of clearances on performance and priming capabilities are virtually unknown to users. Progressing cavity pumps, now widely used in wastewater treatment plants and paper mills, are virtually uncovered in the available literature, and even the principle of their operation is only understood by a few specialists among the designers. The same applies to multiple-screw pumps: a controversy still exists about whether outside screws in three-screw designs provide additional pumping or not.

An example of published literature which when used alone is no longer adequate is A.J. Stepanoff's well-known book *Centrifugal and Axial Flow Pumps*. It describes the theory of centrifugal pumps well, but has no information on actual applications to guide the user and help with actual pump selection for his or her applications. Besides, the material in the book does not nearly cover any of the latest developments, research findings, and field experience in the last 20 to 30 years. Another example comes from a very obscure publication on progressing cavity pumps, *The Progressing Cavity Pumps,* by H. Cholet,[21] published in 1996. However, this book concentrates mostly on downhole applications, and is more of a general overview, with some applicational illustrations, and does not contain any troubleshooting techniques of a "what-to-do-if." In the U.S., this book is essentially unknown and can be obtained only in certain specialized conferences in Europe. There is a good publication by H.P. Bloch, *Process Plant Machinery,*[19] which covers a variety of rotating and stationary machinery, as well as being a good source for the technical professional. It provides an overview of pumps, but for detailed design and applicational specifics, a dedicated book on pumps would be a very good supplement. Finally, the *Kirk-Othmer Encyclopedia of Chemical Technology* contains a chapter on "Pumps," written by the author,[1] and includes comparative descriptions of various pump types, with applicational recommendations and an extensive list of references. However, while being a good reference source, it is generally used primarily as it was intended— as an encyclopedial material, designed to provide the reader with a starting foundation, but is not a substitute for an in-depth publication on pumping details.

For the above reasons, this new book on centrifugal and rotary pumps will provide much needed and timely material to many plant engineers, maintenance personnel, and operators, as well as serving as a relevant textbook for college courses on rotating machinery, which are becoming more and more popular, as technological trends bring the need to study pumping methods to the attention of college curricula. This book is unique not only because it covers the latest pump designs and theory, but also because it provides an unintimidating reference resource to practicing professionals in the U.S. and throughout the world.

# Author

Lev Nelik is Vice President of Engineering and Quality Assurance of Roper Pump Company, located in Commerce, GA. He has 20 years of experience working with centrifugal and positive displacement pumps at Ingersoll-Rand (Ingersoll-Dresser), Goulds Pumps (ITT), and Roper Industries. Dr. Nelik is the Advisory Committee member for the Texas A&M International Pump Users Symposium, an Advisory Board member of *Pumps & Systems Magazine* and *Pumping Technology Magazine*, and a former Associate Technical Editor of the *Journal of Fluids Engineering*. He is a Full Member of the ASME, and a Certified APICS (CIRM). A graduate of Lehigh University with a Ph.D. in Mechanical Engineering and a Masters in Manufacturing Systems, Dr. Nelik is a Registered Professional Engineer, who has published over 40 documents on pumps and related equipment worldwide, including a "Pumps" section for the *Kirk-Othmer Encyclopedia of Chemical Technology* and a section for *The Handbook of Fluid Dynamics* (CRC Press). He consults on pump designs, teaches training courses, and troubleshoots pump equipment and pumping systems applications.

# Acknowledgments

The author wishes to thank people and organizations whose help made this publication possible. Particularly helpful contributions in certain areas of this book were made by:

Mr. John Purcell, Roper Pump: "Gear Pumps"
Mr. Jim Brennan, IMO Pump: "Multiple-Screw Pumps"
Mr. Kent Whitmire, Roper Pump: "Progressing Cavity Pumps"
Mr. Herbert Werner, Fluid Metering, Inc.: "Metering Pumps"
Mr. Luis Rizo, GE Silicones: General feedback as a pump user as well as other comments and assistance which took place during numerous discussions.

Special appreciation for their guidance and assistance goes to the staff of CRC Press, who made this publication possible, as well as thanks for their editorial efforts with text and illustrations, which made this book more presentable and appealing to the readers.

Finally, and with great love, my thanks to my wife, *Helaine*, for putting up with my many hours at home working on pumps instead of on the lawn mower, and to *Adam*, *Asher*, and *Joshua*, for being motivators to their parents.

**Lev Nelik**

# Table of Contents

# 1 Introduction

Pumps are used in a wide range of industrial and residential applications. Pumping equipment is extremely diverse, varying in type, size, and materials of construction. There have been significant new developments in the area of pumping equipment since the early 1980s.[1] There are materials for corrosive applications, modern sealing techniques, improved dry-running capabilities of sealless pumps (that are magnetically driven or canned motor types), and applications of magnetic bearings in multistage high energy pumps. The passage of the Clean Air Act of 1980 by the U.S. Congress, a heightened attention to a safe workplace environment, and users' demand for greater equipment reliability have all led to improved mean time between failures (MTBF) and scheduled maintenance (MTBSM).

# 2 Classification of Pumps

One general source of pump terminology, definitions, rules, and standards is the Hydraulic Institute (HI) Standards,[2] approved by the American National Standards Institute (ANSI) as national standards. A classification of pumps by type, as defined by the HI, is shown in Figure 1.

Pumps are divided into two fundamental types based on the manner in which they transmit energy to the pumped media: kinetic or positive displacement. In kinetic displacement, a centrifugal force of the rotating element, called an impeller, "impels" kinetic energy to the fluid, moving the fluid from pump suction to the discharge. On the other hand, positive displacement uses the reciprocating action of one or several pistons, or a squeezing action of meshing gears, lobes, or other moving bodies, to displace the media from one area into another (i.e., moving the material from suction to discharge). Sometimes the terms 'inlet' (for suction) and 'exit' or 'outlet' (for discharge) are used. The pumped medium is usually liquid; however, many designs can handle solids in the forms of suspension, entrained or dissolved gas, paper pulp, mud, slurries, tars, and other exotic substances, that, at least by appearance, do not resemble liquids. Nevertheless, an overall liquid behavior must be exhibited by the medium in order to be pumped. In other words, the medium must have negligible resistance to tensile stresses.

The HI classifies pumps by type, not by application. The user, however, must ultimately deal with specific applications. Often, based on personal experience, preference for a particular type of pump develops, and this preference is passed on in the particular industry. For example, boiler feed pumps are usually of a multistage diffuser barrel type, especially for the medium and high energy (over 1000 hp) applications, although volute pumps in single or multistage configurations, with radially or axially split casings, also have been applied successfully. Examples of pump types and applications and the reasons behind applicational preferences will follow.

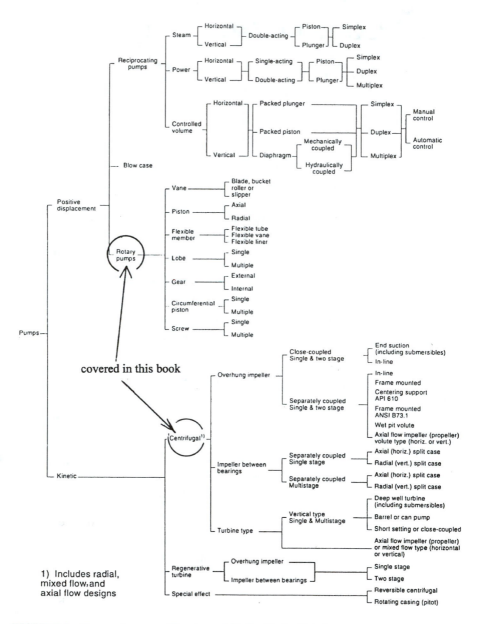

**FIGURE 1**   Types of pumps. (Courtesy of Hydraulic Institute.)

# 3 Concept of a Pumping System

## LIQUID TRANSFER

To truly understand pump operation, one needs to carefully examine the specifics of each individual system in which a pump is installed and operating (see Figure 2). The main elements of a pumping system are:

- Supply side (suction or inlet side)
- Pump (with a driver)
- Delivery side (discharge or process)

The energy delivered to a pump by the driver is spent on useful energy to move the fluid and to overcome losses:

$$\text{Energy}_{input} = \text{Energy}_{useful} + \text{Losses} \qquad (1)$$

$$\text{Efficiency} = \text{Energy}_{useful} / \text{Energy}_{input} \qquad (2)$$

$$\text{Losses} = \text{Mechanical} + \text{Volumetric} + \text{Hydraulic} \qquad (3)$$

| Mechanical | Volumetric | Hydraulic |
|---|---|---|
| ⇓ | ⇓ | ⇓ |
| bearings | leakage (slip) | friction |
| coupling | | entrance/exit |
| rubbing | | vortices |
| | | separation |
| | | disc friction |

From the pump user viewpoint, there are two major parameters of interest:

### *Flow* and *Pressure*

**Flow** is a parameter that tells us *how much of the fluid* needs to be moved (i.e., transferring from a large storage tank to smaller drums for distribution and sale, adding chemicals to a process, etc.).

**Pressure** tells us *how much of the hydraulic resistance* needs to be overcome by the pumping element, in order to move the fluid.

In a perfect world of zero losses, all of the input power would go into moving the flow against given pressure. We could say that all of the available driver power was spent on, or transferred to, a hydraulic (i.e., useful) power. Consider the simple

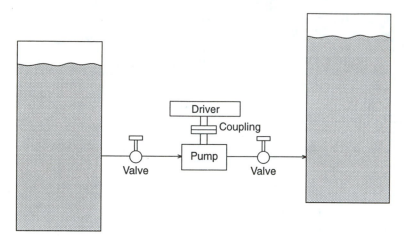

**FIGURE 2**  Pump in a system.

illustration in Figure 3, which shows a piston steadily pushed against pressure, "p," inside a pipe filled with liquid. During the time "t," the piston will travel a distance "L," and the person, exerting force "F" on a piston, is doing work to get this process going. From our school days, we remember that *work* equals *force* multiplied by *distance*:

$$W = F \times L \qquad (4)$$

For a steady motion, the force is balanced by the pressure "p," acting on area, "A":

$$W = (p \times A) \times L = p \times (A \times L) = p \times V \qquad (5)$$

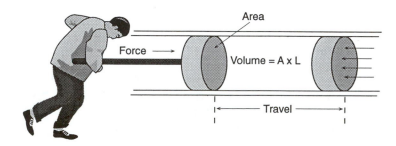

**FIGURE 3**  Concept of power transfer to the fluid.

## INPUT POWER, LOSSES, AND EFFICIENCY

Work per unit of time equals power. So, dividing both sides of the equation by "t," we get:

$$\frac{W}{t} = \frac{p \times V}{t}, \tag{6}$$

or,

$$Power = p \times Q,$$

where

$$Q = \frac{V}{t}.$$

"Q" is the volume per unit of time, which in pump language is called "flow," "capacity," or "delivery." Inside the pump, the fluid is moved against the pressure by a piston, rotary gear, or impeller, etc. (thus far assuming no losses).

This book will use conventional U.S. nomenclature, which can easily be converted to metric units using the conversion formulas located in Appendix B at the end of the book.

So, *Ideal Power = Fluid Horsepower = FHP = p × Q × constant,* since all power goes to "fluid horsepower," in the ideal world. Typically, in U.S. units, pressure is measured in *psi*, and flow in *gpm*, so we derive the constant:

$$psi \times gpm = \frac{lbf \times gal}{in^2 \times min} \times \left(144 \frac{in^2}{ft^2}\right) \times \left(\frac{ft^3}{7.48 \text{ gal}}\right) \times \left(\frac{min}{60 \text{ sec}}\right)$$

$$= \frac{lbf \times ft}{sec} \times \left(\frac{144}{7.48 \times 60}\right) \times \left(\frac{HP}{550 \frac{lbf \times ft}{sec}}\right) = \frac{BHP}{1714}.$$

Therefore,

$$FHP = \frac{p \times Q}{1714}. \tag{7}$$

This is why the "1714" constant "rings a bell" for rotary pump users and manufacturers.

Returning to the "real world," let us "turn on the friction" exerted by the walls of the imperfect pipes on liquid, and consider the rubbing of the piston against the pipe walls, as well as the "sneaking" of some of the liquid back to low pressure through the clearances between the piston and pipe walls. BHP = FHP + Losses, or introducing the efficiency concept:

$$\eta = \frac{FHP}{BHP},$$
(8)

or

$$FHP = BHP \times \eta.$$

We can now correct Equation 7 with the efficiency:

$$BHP = \frac{p \times Q}{\eta \times 1714}.$$
(9)

Jumping ahead a little, Equation 9 is typically used when dealing with positive displacement pumps (which include rotary pumps), but a "centrifugal world" is more accustomed to expressing *pressure* traditionally in feet of head, using specific gravity[3]:

$$H = \frac{p \times 2.31}{SG} (\text{feet of water})$$
(10)

which turns Equation 9 into

$$BHP = \frac{\left(\dfrac{H \times SG}{2.31}\right) \times Q}{\eta \times 1714} = \frac{H \times Q \times SG}{\eta \times 3960}.$$
(11)

This is why a "3960" constant should now "ring a bell" for centrifugal pumps users. Both Equation 9 and 11 produce identical results, providing that proper units are used.

## SYSTEM CURVE

From the discussion above, we have established that *flow* and *pressure* are the two main parameters for a given application. Other parameters, such as pump speed, fluid viscosity, specific gravity, and so on, will have an effect on flow and/or pressure, by modifying the hydraulics of a pumping system in which a given pump operates. A mechanism of such changes can be traced directly to one of the components of losses, namely the hydraulic losses.

Essentially, any flow restriction requires a pressure gradient to overcome it. These restrictions are valves, orifices, turns, and pipe friction. From the fundamentals of hydraulics based on the Bernoulli equation, a pressure drop (i.e., hydraulic loss) is proportional to velocity head:

$$h_{loss} = K \frac{V^2}{2g} \quad \text{(coefficient "k" can be found in books on hydraulics).[3]} \quad (12)$$

For the flow of liquid through a duct (such as pipe), the velocity is equal to:

$$V = \frac{Q}{A} \qquad (13)$$

which means that pressure loss is proportional to the square of flow:

$$h_{loss} \sim Q^2. \qquad (14)$$

If this equation is plotted, it will be a parabola (see Figure 4).

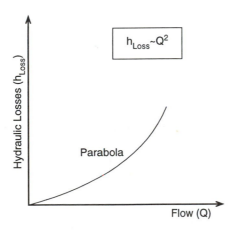

**FIGURE 4**   Hydraulic losses, as a function of flow.

## PUMP CURVE

A pump curve shows a relationship between its two main parameters: *flow* and *pressure*. The shape of this curve (see Figure 5) depends on the particular pump type.

Later on, we will show how these curves are derived. For now, it is important to understand that the energy supplied to a pump (and from a pump to fluid) must overcome a system resistance: mechanical, volumetric, and hydraulic losses. In terms of pressure drop across the pump, it must be equal to the system resistance, or demonstrated mathematically,

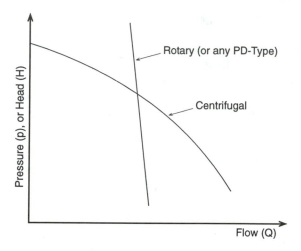

**FIGURE 5** Pump curves, relating pressure and flow. The slope of the centrifugal pump curve is "mostly" flat or horizontal; the slope of the PD-pump is almost a vertical line.

$$\Delta p_{pump} = h_{loss}, \text{ at a given flow.} \tag{15}$$

Therefore, the pump operating point is an intersection of the pump curve and a system curve (see Figure 6). In addition to friction, a pump must also overcome the elevation difference between fluid levels in the discharge and suction side tanks, a so-called static head, that is independent of flow (see Figure 7). If pressure inside the tanks is not equal to atmospheric pressure then the static head must be calculated as equivalent difference between total static pressures (expressed in feet of head) at the pump discharge and suction, usually referenced to the pump centerline (see Figure 8). The above discussion assumes that the suction and discharge piping near the pump flanges are of the same diameter, resulting in the same velocities. In reality, suction and discharge pipe diameters are different (typically, a discharge pipe diameter is smaller). This results in difference between suction and discharge velocities, and their energies (velocity heads) must be accounted for. Therefore, a total pump head is the difference between all three components of the discharge and suction fluid *energy per unit mass*: static pressure heads, velocity heads, and elevations. For example,

$$H = \frac{P_d - P_s}{\gamma} + \frac{V_d^2 - V_s^2}{2g} + (z_d - z_s).^1 \tag{16}$$

Note that the units in Equation 16 are feet of head of water. The conversion between pressure and head is:

$$H = \frac{p \times 2.31}{SG}. \tag{17}$$

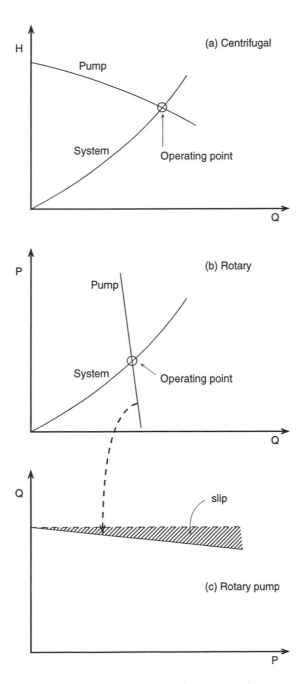

**FIGURE 6** Pump operating point — intersection of a pump and a system curves.
*Note*: Due to the almost vertical curve slope of rotary pumps (b), their performance curves are usually and historically plotted as shown on (c) (i.e., flow vs. pressure).

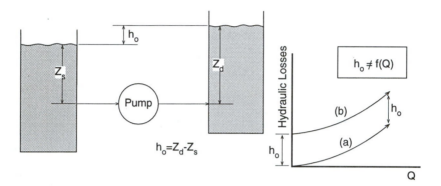

**FIGURE 7**  System curves:
(a)  without static head ($h_o$ = negligible)
(b)  with static head

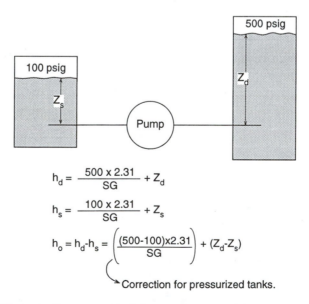

$$h_d = \frac{500 \times 2.31}{SG} + Z_d$$

$$h_s = \frac{100 \times 2.31}{SG} + Z_s$$

$$h_o = h_d - h_s = \left( \frac{(500-100) \times 2.31}{SG} \right) + (Z_d - Z_s)$$

Correction for pressurized tanks.

**FIGURE 8**  "Equivalent" static head, ($h_o$), must be corrected to account for the actual pressure values at the surfaces of fluids in tanks.

From our high school days and basic hydraulics, we remember that the pressure, exerted by a column of water of height, "h," is

$$p = \rho g h = \gamma h, \tag{18}$$

where $\gamma$ is a specific weight of the substance, measured in lbf/ft$^3$. A *specific gravity* (SG) is defined as a ratio of the specific weight of the substance to the specific weight of cold water: $\gamma_o = 62.4$ lbf/ft$^3$. (SG is also equal to the ratio of densities, due to a gravitational constant between the specific weight and density). So,

$$SG = \rho/\rho_o = \gamma/\gamma_o, \tag{19}$$

$$p = \rho g h = \gamma h = (\gamma_o SG)h = 62.4 \times SG \times h \ (lbf/ft^2) \tag{20}$$

(To obtain pressure in more often used units of lbf/in$^2$ (psi), divide by 144).

$$p = \frac{h \times SG}{2.31}, \tag{21}$$

or

$$h = \frac{p \times 2.31}{SG}$$

Clearly, if the system resistance changes, such as an opening or a closing of the discharge valve, or increased friction due to smaller or longer piping, the slope of the system curve will change (see Figure 9). The operating point moves: $1 \rightarrow 2$, as valve becomes "more closed," or $1 \rightarrow 3$, if it opens more.

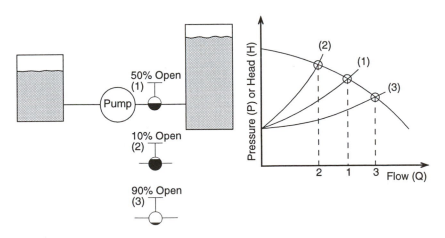

**FIGURE 9**   System curves at different resistance.

# 4 Centrifugal Pumps — Fundamentals

A centrifugal pump is known to be a "pressure generator," vs. a "flow generator," which a rotary pump is. Essentially, a centrifugal pump has a rotating element, or several of them, which "impel" (hence the name impeller) the energy to the fluid. A collector (volute or a diffusor) guides the fluid to discharge. Figure 10 illustrates the principle of the developed head by the centrifugal pump. A good detailed derivation of the ideal head, generated by the impeller, is based on the change of the angular momentum between the impeller inlet and exit.[4] Equation 22 is a final result:

$$H_i = \frac{(V_\theta U)_2 - (V_\theta U)_1}{g} \tag{22}$$

Above, index "1" indicates conditions at the impeller inlet, and index "2" indicates conditions at the impeller exit. The velocity triangles, used to calculate the developed head, must actually be constructed immediately *before* the impeller inlet, and immediately *after* the exit (i.e., slightly outside the impeller itself). The inlet component $(V \times U)_1$ is called pre-rotation, and must be accounted for. In many cases the pre-rotation is zero, as flow enters the impeller in a straight, non-rotating manner. Its effect is relatively small, and we will disregard it in this writing. As flow enters the impeller, the blade row takes over the direction of flow, causing sudden change at the inlet (shock). As flow progresses through the impeller passages, it is guided by the blades in the direction determined by the blade relative angle ($\beta b_2$). However, a flow deviation from the blades occurs and depends on the hydraulic loading of the blades. Parameters affecting this loading include the number of blades and the blade angle. As a result, by the time the flow reaches the impeller exit, its relative direction (flow angle $\beta f_2$) is less than the impeller blade angle ($\beta b_2$). This means that the actual tangential component of the absolute velocity ($V\theta_2$) is less than it would be if constructed solely based on the impeller exit blade angle. The resultant ideal head would, correspondingly, be less. The flow deviation from the blade direction has nothing to do with hydraulic losses, which must be further subtracted from the ideal head, to finally arrive at the actual head ($H_a$).

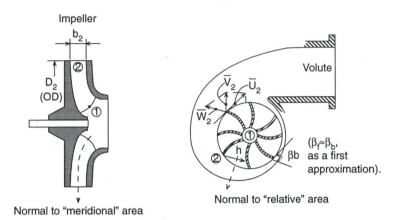

Normal to "meridional" area

Normal to "relative" area

**FIGURE 10A**   $A_{m_2} = \pi D_2 b_2$, $A_x = b_z h_z z$
$z =$ number of blades (8 here)
$\bar{u} =$ peripheral velocity vector
$\bar{w} =$ relative velocity vector
$\bar{v} =$ absolute (resultant) velocity vector
"1" = inlet, "2" = exit of impeller

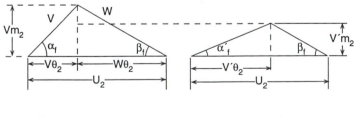

(a) More flow $(V_{m_2} \times A_{m_2})$
Less head $(V_{\theta_2} U_2)$
$V_{m_2} > V'_{m_2}$,

(b) Less flow $(V'_{m_2} A_{m_2})$
More head $(V'_{\theta_2} U_2)$
$V_{\theta_2} < V'_{\theta_2}$

**FIGURE 10B** Impeller velocity triangles.

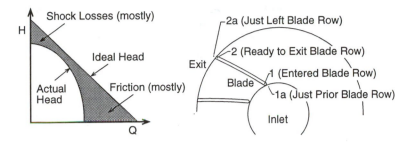

## Entrance:

1a: $V_{1a}$ ↑ velocity vector in absolute direction (relative direction not meaningful yet), $V_{1a} \approx Vm_1$

1: blades dictate direction according to blade inlet angle $\beta_{b1}$

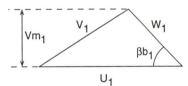

If 1a and 1 represented together

## Exit:

Velocity triangles do not differ very much between 2 and 2a; i.e., there is no shock losses, as compared to inlet. (Actually, they do differ per Busemann's slip factor, but that discussion is significantly beyond the scope and simplifications of this book.)

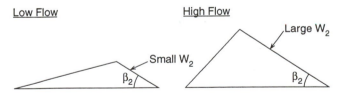

**FIGURE 11**   The nature of hydraulic losses in the impeller.

With regard to flow, it is a product of corresponding velocity and area:

$$Q = V_m \times A_m = W \times A_x. \tag{23}$$

"Vm" is called a "meriodional" velocity (a component of the absolute velocity into the meriodional direction), and "W" is a relative velocity of the fluid passing through the impeller passages. Hydraulic losses reduce the generated head, as shown in Figure 11. Note that at low flow the relative velocity (W) through the impeller is small. Since relative velocity characterizes the movement of fluid through the impeller passages, it determines friction losses: relative velocity is higher at higher flow, hence friction losses are predominant there. At low flow, the relative velocity is small and friction is negligible. The main component of losses is inlet recirculation and incidence. At design point (BEP) the fluid enters the impeller smoothly (shockless), which can be seen as the absolute velocity $V_1$ is presented as a vectorial triangle that includes $V_1$, $U_1$, and $W_1$. Therefore, on both sides of the best efficiency point (BEP), the relative component $W_{1a}$ (just before entering the blades row) would be different from $W_1$ (just inside), causing shock losses. As a result, at higher flow, friction is a predominant loss component, and at low flow, incidence and shock are predominant loss components.

## AFFINITY LAWS

In an Affinity Law, the ratios between the internal hydraulic parameters of a given device, such as a pump, remain constant when an "external" influence is exerted on the device. Rotating speed (RPM) is one such influence. When pump speed changes, the "affinity" of the velocity triangles (i.e., their shape) remains the same (see Figure 12).

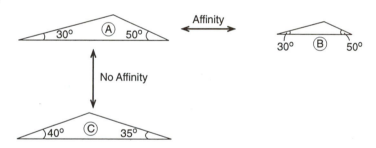

**FIGURE 12** Illustration of the Affinity principle.

In geometric terms, this means that the *relative flow angle, $\beta_f$* (which is, as first approximation could be assumed, equal to the *impeller blade angle, $\beta_b$*), as well as the *absolute flow angle* $\alpha_f$ all remain constant as the speed increases (see Figure 13).

The ratio of velocities also remains constant:

$$\frac{U_A}{U_B} = \frac{V_{mA}}{V_{mB}} = \frac{W_A}{W_B} = \frac{V_{\theta A}}{V_{\theta B}} = \frac{V_A}{V_B} = \frac{RPM_A}{RPM_B}. \tag{24}$$

Since velocity is proportional to RPM, and flow is in direct relationship with velocity (see Equation 9), the flow is thus proportional to RPM. From Equation 8, the ideal head is proportional to $(V_\theta U)_2$ (i.e., $H_i \sim RPM^2$), and, neglecting a correction for losses, the actual head is therefore approximately proportional to RPM squared, $H \sim RPM^2$.

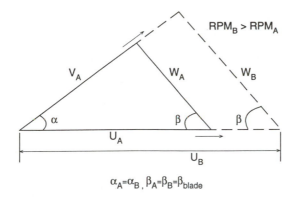

$$\alpha_A = \alpha_B, \ \beta_A = \beta_B = \beta_{blade}$$

**FIGURE 13** Affinity is preserved as rotational speed changes from (A) to (B).

Finally, the power which is a product of head and flow (Equation 6) is proportional to the cube of RPM. To summarize:

$$Q \sim RPM \tag{25a}$$

$$H \sim RPM^2 \tag{25b}$$

$$BHP \sim RPM^3. \tag{25c}$$

## HELPFUL FORMULAS, PER CENTRIFUGAL PUMP TRIANGLES

$$\frac{V_m}{W} = \sin\beta, \quad V_m = W\sin\beta \tag{26}$$

$$U - V_\theta = W_\theta = W\cos\beta, \quad V_\theta = U - W\cos\beta \tag{27}$$

Neglecting the inlet "pre-whirl" component $(V_\theta U)_1$:

$$H_i = \left(\frac{V_\theta U}{g}\right)_2 = \frac{(U - W\cos\beta)_2 U_2}{g} = \frac{\left(U - \dfrac{Q\cos\beta}{Ax}\right)_2 U_2}{g} \tag{28}$$

$$= \frac{u_2^2}{g} - \frac{QU_2\cos\beta_2}{A_x g}.$$

From the previous formula for the ideal head, a graphical representation (straight line) can be constructed, as shown in Figure 14.

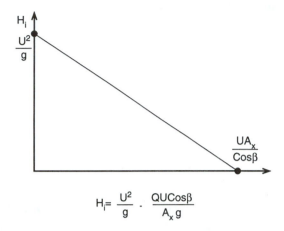

$$H_i = \frac{U^2}{g} - \frac{QUCos\beta}{A_x g}$$

**FIGURE 14** Construction of the ideal head vs. flow.

$$\text{At } Q = 0,\ H_i = \frac{U_2^2}{g}.$$

$$\text{At } H = 0,\ Q = \frac{U_2 A_x}{\cos\beta_2}.$$

$$\left( U_2 = \frac{RPM \times OD}{229} \text{ and } V_{m_2} = \frac{Q \times 0.321}{A_{m_2}} \right).$$

Note: Linear dimensions are in inches, area is in square inches, flow is in gpm, head is in feet, and velocities are in ft/sec.

## QUIZ #1 — VELOCITY TRIANGLES

Referring to Figure 15, construct a velocity triangle at the impeller exit, and predict a pump head for this 50 gpm single-stage overhung centrifugal pump, running at 1800 RPM. What is an absolute flow angle $\alpha_{f2}$?

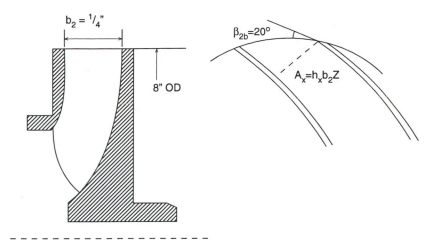

**FIGURE 15** Impeller geometry for Quiz #1.

## Solution to Quiz #1

$$U_2 = \frac{RPM \times D_2}{229} = \frac{1800 \times 8}{229} = 63 \text{ ft/sec}$$

$$A_{m_2} = \pi D_2 b_2 = 3.14 \times 8 \times 0.25 = 6.3 \text{ in}^2$$

$$V_{m_2} = \frac{Q \times 0.321}{A_{m_2}} = \frac{50 \times 0.321}{6.3} = 2.6 \text{ ft/sec}$$

$$W_2 = \frac{V_{m_2}}{\sin \beta_2} = \frac{2.6}{\sin 20°} = 7.5 \text{ ft/sec}$$

$$V_{\theta_2} = U_2 - W_2 \cos \beta_2 = 63 - 7.5 \times \cos 20° = 56 \text{ ft/sec}$$

$$H_i = \frac{V_{\theta_2} U_2}{g} = \frac{56 \times 63}{32.2} = 110 \text{ ft}$$

As was mentioned earlier, we are using an impeller exit angle for the exit velocity triangles, disregarding flow angle deviation from the blade angle, as an approximation (i.e., assuming $= \beta_2 = \beta_{f2} \approx \beta_{b2}$).

$$\alpha_{f_2} = \tan^{-1} \frac{V_{m_2}}{V_{\theta_2}} = \tan^{-1} \frac{2.6}{56} = 2.7°$$

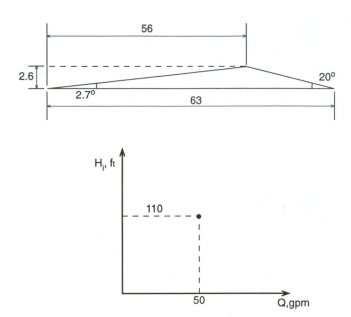

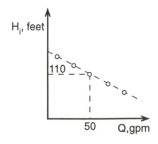

**FIGURE 16** Resultant impeller exit velocity triangle and operating point for Quiz #1.

The resultant velocity triangle and a single operating point on the performance curve are shown in Figure 16. To construct a complete curve, such calculations must be performed for several flows, producing a series of points of the ideal head (Figure 17). If we could now calculate hydraulic losses at each flow, we would obtain another set of points for the actual head, which is the next step (see Figure 18).

**FIGURE 17** Ideal head is calculated at several flows, producing a set of $H_i$-Q points.

However, these calculations are too involved for the scope of this book, and the composition of losses is very different for different pump types. An approximate technique could be used for a rough estimate of the H-Q curve, as explained below.

The 50 gpm in previous discussion was an arbitrary flow. If the 50 gpm is actually a best efficiency point (BEP) of this pump, then as a first approximation and as a rule of thumb, a value of 85% for the hydraulic efficiency can be used at BEP, and is typical (practical) for many API and ANSI type centrifugal pumps. The

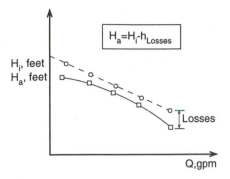

**FIGURE 18** Calculated losses, at each flow, are subtracted from the ideal head ($H_i$), producing a new set of points for the actual head ($H_a$).

value of hydraulic efficiency at off-BEP points is a more involved matter. The actual head at BEP would then be equal to:

$$H = H_i \times \eta_H = 110 \times 0.85 = 94 \text{ ft .}$$

The actual head at zero flow (shutoff), for these types of pumps, ranges between 10 to 30% higher than the BEP head, which means:

$$H_{so} \approx 1.2 \times 94 \text{ ft} = 113 \text{ ft .}$$

Having these two points, we can roughly sketch the approximate the H-Q curve (Figure 19).

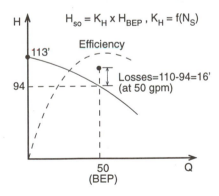

**FIGURE 19** Shut-off (shut valve) head, $H_{so}$, is assumed as 20% rise from the BEP value, which is typical for many pumps, in a wide range of specific speeds ($N_s$) is discussed later.

Let us also construct a system curve. Suppose that the pump in Quiz #1 was installed in a system, as shown in Figure 2, with relatively short piping, and a 60 ft differential between the liquid levels in its tanks (i.e., 60 ft static head). For relatively low viscosity liquids (applications in which centrifugal pumps mostly are used), friction losses can be neglected for a short piping run, and therefore all pressure drop (hydraulic loss) is taken across the valve — a resultant parabola, as was explained earlier, is shown in Figure 20.

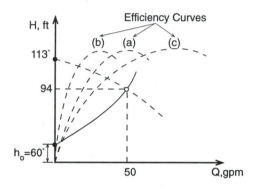

*Note 1*: Thus far only two points of pump curve have been calculated (at 50 gpm and at shut-off).

*Note 2*: Efficiency curves demonstrate possible positions (depending on pump internals) of operating point (a) if 50 gpm is BEP (assumed here), and (b) and (c) are for BEP to the left of 50 gpm, or to the right of it.

**FIGURE 20** Operating point — intersection of pump and system curves.

$$H_{valve} = H_{static} + KQ^2 \tag{30}$$

Substituting the known values from the pump data in the previous valve equation, we can get the coefficient "k":

$$94' = 60' + K \times 50^2, K = 0.014.$$

This finalizes the valve (i.e., system) equation,

$$H_{system} = 60' + 0.014Q^2,$$

which can be re-plotted now more accurately in Figure 20.

## PERFORMANCE CURVES

Mathematically, if the discharge valve is throttled, its loss coefficient "k" changes (higher for a more closed valve). Inside the pump, however, there is one particular flow (BEP), where the hydraulic losses are minimal. Generally, at higher flows, the friction losses are predominant, and at lower flows the more significant components of losses are flow separation and vortices (see Figure 11). The flow requirements in many applications change continuously: the production requirements change, different liquids require different amounts of additives to the process, etc. A common and simple way to change the flow is to open or close the discharge valve; however, this method is also the least efficient. As we shall see later, for a type of pump in Quiz #1, a 20% reduction in flow may cause a 10 to 12% loss in efficiency, and this costs money.

## QUIZ #2 — HOW MUCH MONEY DID A MAINTENANCE MECHANIC SAVE HIS PLANT?

Refer to Figure 21: How much money would a mechanic in a chemical plant bring home next year, due to a raise he got for finding a better way to change the pump flow than by valving? (There is no need to assume 50 gpm as a BEP point; it can be anywhere along the performance curve, since the example is only for relative comparative illustration.)

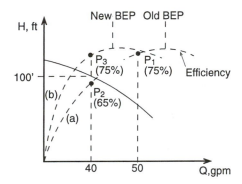

**FIGURE 21** Different ways to shift pump operating point, for Quiz #2.

(a) *Inefficient way:* Pump operating point shifted from ($P_1$) 50 gpm to 40 gpm ($P_2$), by valve throttling, with efficiency drop from 75 to 65%.
(b) *Efficient way:* Operating point moved from $P_1$ to $P_3$, by volute modifications (or, equally effective, by speed change with VFD), keeping efficiency high at 75%.

## Solution to Quiz #2

1. $\text{BHP}_{\text{(inefficient way)}} = \dfrac{H \times Q \times SG}{3960\eta} = \dfrac{40 \times 100 \times 1.0}{3960 \times 0.65} = 1.55 \text{ HP}$

2. $\text{BHP}_{\text{(efficient way)}} = \dfrac{40 \times 100}{3960 \times 0.75} = 1.35 \text{ HP}$

3. $\Delta = 0.2 \text{ HP} \approx 0.15 \text{ kW}$

For one year (8760 hours), at 10 cents per kilowatt, this would result in:

$$(0.15 \times 8760) \times 0.10 = \$130 \text{ per pump}$$

An example of such an efficient way would be a volute (or diffusor) modification, to rematch it to the new operating conditions of 40 gpm. Another way is to reduce the speed, which would reduce the flow proportionally — this can be done with a variable frequency drive (VFD). For a 1000-pump plant, this could be a savings of $130,000. Now, if the mechanic's boss would allocate at least 1% of savings to the mechanic's bonus, this $1300 would make a nice present at the year's end! And this is just for a small 1.5 HP pump — how about the larger ones?!

## PERFORMANCE MODIFICATIONS

Let us now assume that the pump in our previous example has been properly sized for the application, and has been operating at its best efficiency point, Q = 50 gpm, and developing head, H = 94 ft — the operating point being an intersection of a pump curve and a system curve, with static head of h = 60 ft. The pump's suction line is connected to the supply tank, which is positioned at the supply silo, below the ground level. A supplier of chemicals used by the processing plant has advised that the raw chemicals can be supplied in truck-mounted containers and pumped out directly from the trucks, without unloading of the containers to the ground, thereby saving time and money. As a result, a suction level is now higher, and, since nothing different was done on the discharge side, the net static level (discharge level minus suction level) has decreased from the original 60 ft to 20 ft. All other variables, including the valve setting, remain the same.

A new system curve move, is shown in Figure 22, and intersects the pump curve at a higher flow (65 gpm) and a lower head (80 ft). The pump now operates at the run-out, to the right of the BEP, and the efficiency dropped from its BEP value of 75% to 68%, as shown on the efficiency curve.

The equation for the system changed only with regard to a static head is:

$$H_{system} = 20 + 0.014 \ Q^2.$$

(i.e., 60' changed to 20', but the valve constant k = 0.014 remained the same, for the same valve setting).

A few weeks later, a plant started to process a new chemical, requiring only 30 gpm flow. A discharge valve was closed, forcing the pump back along its curve. At 30 gpm, the pump is developing 105 feet of head, and is operating at 60% efficiency (see Figure 23). The higher value of head is not important, and does not affect the process. It is anticipated that the plant will continue to produce the new chemical for a long time, and a plant manager had expressed concern about the pump operating at low efficiency. Operators also noticed that other problems with the pump had started to crop up: higher vibrations, seal failures, and noise, emanating from somewhere near the suction end of a pump, which sounded like cavitation. The issue became to maintain 30 gpm flow, but in such a way that the pump would still operate at the BEP. This challenge is similar to the problem in Quiz #2.

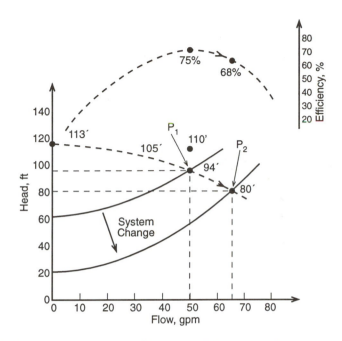

**FIGURE 22** $P_1$: original operating conditions (Q = 50 gpm, H = 94')
$P_2$: system static head changed from 60' to 20', causing the pump to run out to
Q = 65 gpm, H = 80') (Note efficiency drop.)

An engineering consultant has recommended to trim the impeller outside diameter (OD), claiming that the pump BEP would move in certain proportion to the impeller cut: flow would move in direct ratio with a cut, and head would change as a square of a cut, similar to the relationship of flow and head with speed. A plant engineer was skeptical about this, asked to see actual calculations based on the first principles using velocity triangles, which he learned from the pump hydraulics course he took the previous year. The consultant made the analysis, and below are his reasons.

Since the discharge valve setting remains the same, the system curve does not change, but the pump curve would change. The new BEP needs to be at 30 gpm, and the system curve shows 45 feet of head (which can be read from the system curve in Figure 24 or calculated from the system curve equation established earlier). Therefore, the impeller OD cut would need to be such that, at 30 gpm, it would produce 45 feet of head. This requires several iterations. As a first iteration, try a 10% cut:

$$8 \times 0.9 = 7.2", \text{ new impeller OD (D}_2\text{)}.$$

Then,

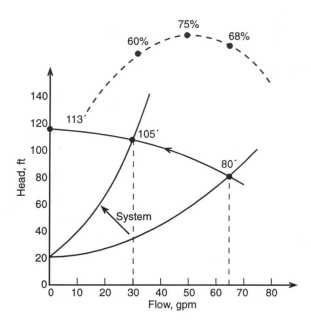

**FIGURE 23** Valve closing caused system curve to intersect pump curve at new point. Pump operation point moved from Q = 65 gpm, H = 80', to Q = 30 gpm, H = 105'. Pump is now operating to the right of BEP, at 60% efficiency.

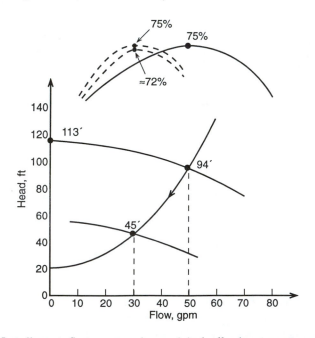

**FIGURE 24** Impeller cut. System curve is as original, allowing pump to operate at BEP, which moves from 50 gpm to 30 gpm efficiently. (Actually, efficiency will show a slight decrease, 2 to 3%.)

$$U_2 = \frac{7.2 \times 1800}{229} = 56.5 \text{ ft/sec},$$

$$A_{m_2} = \pi D_2 b_2 = 3.14 \times 7.2 \times 0.25 = 5.7 \text{ in}^2,$$

$$V_{m_2} = \frac{30 \times 0.321}{5.7} = 1.7 \text{ ft/sec}, \quad V_{\theta_2} = 56.5 - \frac{1.7}{\tan 20°} = 51.9 \text{ ft/sec},$$

$$H_i = \frac{V_{\theta_2} U_2}{g} = \frac{51.9 \times 56.5}{g} = 91 \text{ ft},$$

and

$$H_{\text{actual}} \approx 91 \times 0.85 = 77 \text{ ft} \neq 45 \text{ ft}.$$

A guessed cut is insufficient, since the impeller is generating too much head. Now try a 25% cut:

$$8 \times 0.75 = 6.0" \text{ (new impeller OD)}.$$

Then,

$$U_2 = \frac{6.0 \times 1800}{229} = 47.1 \text{ ft/sec},$$

$$A_{m_2} = \pi \times 6.0 \times 0.25 = 4.7 \text{ in}^2,$$

$$V_{m_2} = \frac{30 \times 0.321}{6.0} = 1.6 \text{ ft/sec},$$

$$V_{\theta_2} = 47.1 - \frac{1.6}{\tan 20°} = 42.7 \text{ ft/sec},$$

$$H_i = \frac{42.7 \times 47.1}{g} = 62 \text{ ft},$$

and

$$H_{\text{actual}} \approx 62 \times 0.85 = 44'.$$

It is now close to the required H = 45'. The remainder of the curve could be constructed similarly, using several flow-points along the performance curve, except the shut-off. The approximate new impeller curve is shown also in Figure 24. The efficiency is preserved at approximately 75%, which made the plant manager happy. In reality, efficiency would drop slightly when BEP moves toward the lower flow (and increase when the opposite is done) toward the higher flow. The actual cut is also exaggerated — 25% is too much and is only used here to illustrate the principle.

Usually, a 10–15% cut is maximum. Beyond that, a pump efficiency begins to suffer appreciably, due to the fact that the impeller "loses" its blades.

The solution worked, but the plant engineer still had an issue with the consultant. He pointed out that the impeller cut was 0.75 (6.0/8.0), the flow changed by 0.6 (30/50), and the head changed by 0.56 (45/94) (i.e., both flow and head changed almost as a square of the cut: ($0.75^2 = .56$)). Yet, the consultant claimed the flow would change linearly. Both were right, for the following reason: theoretically, both flow and head should indeed change as a square of the OD cut, as can be easily seen from the change in velocity triangles. With the change of OD, peripheral velocity changes in direct proportion (see Figure 25), and the tangential component does the same, in order to preserve the relative and absolute angles: maintain BEP, and the hydraulic losses are minimized (matching the flow to the blades as before the cut). The incidence of absolute velocity at the volute tips must also be similar (i.e., maintaining direction). Such preservance of the velocity triangles at the discharge is, indeed, the Affinity Law.

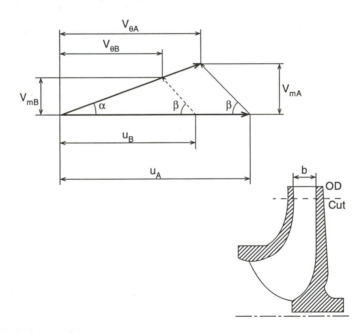

$$f = OD_B/OD_A, \text{ cut}$$
$$U_B = fU_A$$
$$V_{\theta B} = fV_{\theta A}, V_{m_B} = fV_{m_A}$$
$$A_{m_B} = \pi Db = \pi(fD_A)\, b = fA_{m_A}, \text{ if } b = const.$$

$$H_{i_B} = \frac{(V_\theta U)_B}{g} = \frac{f^2(V_\theta U)_B}{g} = f^2 \times H_{i_A}, \quad Q_B = V_{m_B}A_{m_B} = fV_{m_A}fA_{m_A} = f^2Q_A$$

But $f\, b \ne const.$ (e.g., $b_B \approx \dfrac{b_A}{f}$), then: $A_{m_B} = \pi D_B b_B = \pi(fD_A)(\dfrac{b_A}{f}) = A_{m_A}$, and $Q_B = fQ_A$

**FIGURE 25** Effects of the impeller cut.

Therefore, all velocities are reduced linearly with the cut ratio. The meridional area (as well as relative area) changes *linearly* since the diameter is decreased, but the width is not. The product (i.e., flow) of meridional velocity multiplied by the area, therefore changes as a *square* of the cut. This is what the plant engineer had observed. The reason for the consultant's claim that flow is *linear* with the cut is that, *usually*, the impeller width also changes (it gets wider). This change compensates for the *decrease* in the diameter, with the resulting area staying approximately the same. The product of the smaller velocity and constant area would therefore produce the linear flow relationship.

It has been found empirically that the reduction in the impeller diameter tends to follow, in practice, a relationship similar to affinity law for the speed changes:

$$Q \sim OD \tag{31a}$$

$$H \sim OD^2 \tag{31b}$$

$$BHP \sim OD^3. \tag{31c}$$

The above relationships were proposed by A.J. Stepanoff and are valid for a considerable range of specific speeds, except for very high or very low. Impellers of very *low* specific speeds have narrow impeller width, which stays nearly constant for a considerable distance away from the OD. The *high* specific speed designs have short blades, which are more sensitive to a cut, causing them to deviate from the above general rule.

## QUIZ #3 — A VALVE PUZZLE

Why was the valve throttled so much to begin with? (Hint: Did we just use it to illustrate the *mechanism* of flow/pressure restriction?)

**Answer to Quiz #3:**

A valve was used for illustration purposes only, in order to simulate the hydraulic loss. It just as well could have been any equivalent length of pipe friction due to flow of viscous fluid in a long pipe of small diameter. Of course, in a real situation, you should try to keep the valves sized properly, keeping them in the most open position, to avoid unnecessary losses. For flow control purposes and for ease of illustration, a valve can be opened to mathematically reduce system losses, and the pump runs out in flow; but the friction loss cannot be "turned off" (i.e., it would be more difficult to illustrate the concept).

## UNDERFILING AND OVERFILING

**Underfiling** (see Figure 26) is a way to achieve modest (3 to 5%) gain in head. The increased exit area causes a reduction in impeller relative velocity and an increase in tangential component, $V_\theta$.

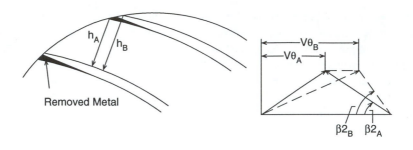

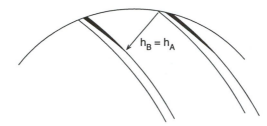

Underfiling: $h_B > h_A$, $\beta2_B > \beta2_A$, $V_{\theta2B} > V_{\theta2A}$, $A_x = h \times b \times z$,

$$A_{x_B} > A_{x_A} \quad H_B = \frac{(V_\theta U)_B}{g} > H_A = \frac{(V_\theta U)_A}{g}$$

Overfiling: The net area between blades does not change, i.e., no appreciable change in head (except slight, due to smoother blades, improved efficiency).

**FIGURE 26** Underfiling and overfiling of blade exits.

$$W = \frac{Q}{A_x}, \quad V_\theta = U - W \times \cos\beta \qquad (32)$$

which causes head increase for a given flow. This also can be viewed as if the exit angle was increased via underfiling. These two views of the same resulting effect reflect different approaches to pump hydraulics by the two traditional schools of thought: Anderson (area method) vs. Stepanoff (angles). Either approach, if followed correctly, leads to the same results.

**Overfiling** is a technique to increase the efficiencies (though modest) by making the exits smoother. Notice that the exit area does not change, thereby having no effect on head-capacity curve. (In actuality, a slight improvement in head does occur due to lowering friction and reduced exit turbulence.)

## DESIGN MODELING TECHNIQUES

When designing a new pump, a designer has two choices: to design from scratch, or to model from other available (similar) designs.

"Blank piece of paper" designs are rare, and are developed for special applications and extreme or unusual conditions. Computer programs on CAD are at the disposal of pump hydraulic design engineers, and are often in the realm of the R&D departments specializing in such activities.

A much more conventional approach is using *modeling techniques*,[4]

$$Q \sim S^3 \tag{33a}$$

$$H \sim S^2 \tag{33b}$$

$$BHP \sim S^5 \tag{33c}$$

where "s" is a linear scaling dimension of all linear dimensions (i.e., multiple of the diameter, width, eye size, etc.). The deviation of Equation 33 is beyond the scope of this book, and is therefore given for reference only.

## SPECIFIC SPEED (Ns)

Pump Specific Speed is defined as a dimensionless parameter, equal to

$$Ns = \frac{RPM \times \sqrt{Q}}{H^{0.75}}, \tag{34}$$

where $Q$ = gpm, and $H$ = feet. The question is sometimes asked, "Why do we call the Ns 'dimensionless,' when in fact, by direct substitution of values, it *does* have a dimensional property?" This property is true and can be shown by straight substitution:

$$\frac{RPM \times \sqrt{gpm}}{ft^{0.75}} = \frac{\left(\dfrac{1}{min}\right) \times \left(\dfrac{gal}{min}\right)^{1/2}}{ft^{3/4}} = \frac{gal^{1/2} \times \left(\dfrac{ft^3}{7.48\ gal}\right)^{1/2}}{min^{3/2} \times ft^{3/4}}$$

$$= (\text{dropping off the constants}) = \frac{ft^{3/2}}{min^{3/4}\ ft^{3/4}} = \frac{ft^{3/4}}{min^{3/2}} \neq unity = 1.$$

However, if a *consistent* set of units is used, including a gravitational constant "g," then a nondimensionality can indeed be demonstrated:

$$\Omega_s = \frac{N\sqrt{Q}}{(gH)^{3/4}} = \frac{\dfrac{1}{sec}\sqrt{ft^3/sec}}{\left(\dfrac{ft}{s^2}\ ft\right)^{3/4}} = \frac{ft^{3/2}}{sec^{3/2}\left(\dfrac{ft}{sec}\right)^{3/2}} = 1 = unity. \tag{35}$$

Here $\Omega_s$ is nondimensional and is called a "universal specific speed." Pump efficiency depends on specific speed and flow, as illustrated in Figure 27.

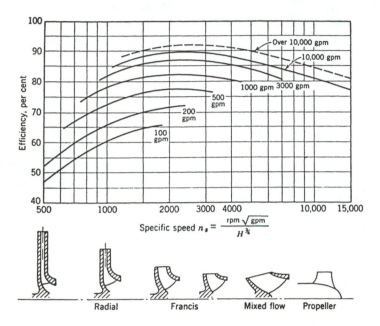

FIGURE 27 Pump efficiency as a function of specific speed and flow.[4]

# 5 Gear Pumps — Fundamentals

Gear pumps belong to a positive displacement rotary group, and are made by enclosing two or more gears in a close-fitting housing. A driver turns a shaft connected to one of the gears, causing it to rotate. This gear drives the other gear through the meshing of the teeth of the two gears, just as with power transmission gears.[5] As the gears rotate, on one side, the teeth are coming out of mesh with each other (see Figure 28). As a tooth is pulled out of the space between two teeth of the other gear, it creates a vacuum. Since the housing forms a seal all around the set of gears, the liquid that rushes into this space to fill this void has to come in through the pump's suction port. Once the spaces between gear teeth are filled with liquid, the liquid rides in these pockets, trapped in place by the housing, until it reaches the discharge side of the pump. The liquid stays in place between the teeth until it reaches the other side of the gear mesh, where the teeth are coming together. Then, when a tooth from the other gear comes into the space between the teeth, the liquid there is forced out. Since the housing still forms a seal around the gears, the only place for the displaced liquid to go is out the pump's discharge port. The pump thus operates like a conveyor belt, with the pockets of liquid between the gear teeth being picked up at the gear mesh, carried to the other side, and dropped off at the other side of the mesh.

There are two basic types of gear pumps: external and internal. External gear pumps usually have two gears with an equal number of teeth on the outside of each gear. Internal gear pumps have one larger gear with the teeth turned inward, meshing with a smaller gear with external teeth. If the larger gear has one tooth more than the inner gear, the two gears form a seal by themselves. If the larger gear has at least two teeth more than the smaller gear, then a crescent-shaped projection of the housing goes between the two gears to help form a seal. The operating principle is the same for all of these types of pumps, and they operate in similar fashion.

The displacement of a pump is the volume of liquid moved in those pockets between gear teeth. It is the theoretical output of the pump before any losses are subtracted. The instantaneous mode of displacement varies slightly as the teeth move through different positions in the mesh, so displacement per shaft revolution cannot be calculated exactly. However, there are some good approximations. For example, if the cavity area between the gear teeth is assumed to be approximately equal to the area of the teeth themselves (i.e., a cavity is an inverse of a tooth), then the

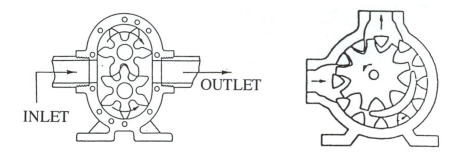

External gear                              Internal gear

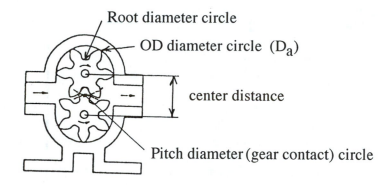

Addendum = radial distance from the pitch circle

and OD circle $\left(a = \dfrac{D_a - D_p}{4}\right)$

Dedendum = radial distance from the pitch circle

to the root circle $\left(b = \dfrac{D_p - D_r}{4}\right)$

**FIGURE 28** Different types of gear pumps, with geometry illustrations.

displacement per revolution would be equal to the volume occupied by half the space between the gear addendum (outside diameter) and a root diameter, multiplied by two (to account for two gears), times gear width:

$$q = \frac{\pi}{4}\left(D_a^2 - D_r^2\right) \times \frac{1}{2} \times 2 \times W\left(\text{in}^3/\text{rev}\right); \tag{36}$$

or, simplifying and dividing by 231, to convert in$^3$/rev to gal/rev:

$$q = 0.0034 \times \left( D_a^2 - D_r^2 \right) W, \tag{37}$$

where

$q$   = displacement (gallons/revolution)
$D_a$ = gear outside diameter (inches)
$D_r$ = gear root diameter (inches)
$W$  = gear face width (inches).

This formula assumes that both gears have the same outside diameter and number of teeth. The addendum of gears for pumps is often extended when compared to power transmission gears. This is to increase the pump's displacement. Gears with smaller numbers of teeth have larger addendum for a given center distance. So, most gear pumps have 12 or less teeth on the gears. Examples of gears with various numbers of teeth, pitches, and sizes are shown in Figure 29.

**FIGURE 29** Pumping gears with different number of teeth (Z), pitch diameters ($D_p$), and pitches (Z/$D_p$).

Some pumps have as few as six teeth. This is about the minimum for relatively smooth power transmission between gears. Lobe pumps are similar to gear pumps with two or three teeth, but they use separate timing gears, outside of the liquid, to transmit power from the driving to the driven shaft.

## QUIZ #4 — GEAR PUMP CAPACITY

A plant mechanic has measured the pump gear at the repair shop:

$$D_a = \text{Gear OD} = 3"$$
$$D_r = \text{ID (root)} = 2"$$
$$\text{Width} = 4"$$

Predict how much oil this gear pump will deliver at 1200 RPM.

### SOLUTION TO QUIZ #4

Using Equation 37,

$$q = 0.0034 \times \left(D_a^2 - D_r^2\right) W = 0.0034\left(3^2 - 2^2\right) \times 4 = 0.07 \text{ gal/rev},$$

$$Q = q \times \text{RPM} = 0.07 \times 1200 = 80 \text{ gpm}.$$

Slip is the difference between the theoretical flow (displacement × speed) and actual flow, assuming that there is no cavitation. Slip is the leakage of liquid from the high pressure side of the pump back to the low pressure side. There are a number of separate slip paths in any gear pump, including any liquid from the outlet of the pump that is bled off to flush a seal chamber or lubricate bearings. Three paths are common to all gear pumps: between the ends of the gears and the endplates (known as lateral clearance), between the tips of the gear teeth and the inside of the casing (known as radial clearance), and between the profiles of the meshing teeth. The slip through this last path is very small and is usually ignored.

Slip varies strongly with differential pressure and viscosity and, to some extent, with speed. Slip is directly proportional to differential pressure. It varies inversely, but not proportionally, with viscosity. Slip varies asymptotically with viscosity, approaching zero slip at high viscosities. This means that at low viscosities, small changes can mean large differences in slip. Slip varies inversely with speed to a small extent, but this is normally ignored, and predictions are made slightly conservatively at higher speeds. There is also a strong relationship between clearances and slip. Slip, through a particular clearance, varies directly with the cube of that clearance. This is similar to the oil flow in a hydrodynamic bearing,[6] and, indeed, the interaction between gear ends and the casing wall (or wearplate) is also similar, producing a like hydrodynamic bearing effect. This means that if you double the lateral clearance, you will get eight times as much slip through that clearance. The percentage of slip through each slip path varies with pump design, but in most gear pumps, over half of the slip goes through the lateral clearance; this is because it is usually the largest clearance and it has the shortest distance from high to low pressure. This is why some high pressure pumps eliminate lateral clearance altogether by using discharge pressure to hold movable endplates against the gear faces while the pump is running.

## CAVITATION IN GEAR PUMPS

Cavitation is the formation of voids or bubbles in a liquid as the pressure drops below the vapor pressure of the liquid in the pump's inlet. These bubbles then collapse when they reach the high pressure side of the pump. This collapse can, over time, damage the pump and erode hard surfaces. Cavitation causes a drop in output flow that can sometimes be mistaken for slip, but cavitation can usually be identified by its distinctive sound. Significant cavitation will usually sound like gravel rattling around inside the pump. A rule of thumb is that the liquid velocity in the inlet port should be no more than 5 ft/sec for low-required net inlet pressure.

When pumping viscous fluids, the rotational speed of the pump must be such that the fluid has enough time to fill the voids between gear teeth at the inlet. In other words, the pump can only move the fluid out if there is sufficient suction pressure to push the liquid into the pump inlet. Otherwise, the voids are not filled completely, effectively reducing actual flow through the pump. Therefore, minimum allowable suction pressure depends on the rotating speed, size (pitch diameter), number of gear teeth, and viscosity of the fluid. An empirical approximate relationship, based on charts is[7]:

$$P_{min} = \frac{V_p^{0.826} \times SSU^{0.09}}{61.4} (psia) \tag{38}$$

where

$$Vp = \frac{D_p \times RPM \times 3.13}{Z} (in/min \ per \ tooth),$$

and

Z   = number of gear teeth
$D_p$  = pitch diameter (inches)
SSU = viscosity.

Therefore, pump suction pressure must be greater than the minimum allowable value. If this condition is not maintained, the pump flow will decrease, accompanied by noise, vibrations, and possible damage to the equipment. The cavitation damage in gear pumps, however, is not as severe as in centrifugal pumps. Typically, gear pumps are used for oil and similar liquids which have a significantly lower cavitation (boiling) intensity. The resulting bubbles implode less vigorously than in the case of cold water, and their impact against the equipment's internal boundaries is, therefore, less severe.

## TRAPPING METHODS

As the teeth mesh, they can form closed spaces where one pair of teeth come together before the pair ahead of them has broken contact. This closed space decreases in volume as the teeth continue to move, and the liquid trapped inside can reach very high pressures and come out through the pump's lateral clearances at extremely high velocities. This damages the pump by eroding the endplates and gears; it also causes excessive noise and power consumption by the pump. Trapping problems are worse at high speeds and high pressures. Two methods are used to prevent trapping: helical gears and grooves.

Helical gears have the teeth twisted around the gear in a helix, so that any two teeth can only be in contact at one point. The trapped liquid can move axially along both teeth to escape the mesh.

Grooves in the gears or endplates are used to provide escape routes for trapped liquid. Grooves in endplates must be open to the trapped volume until it reaches its minimum size, and then another groove opens to the trapped space as it expands again. The two grooves must stay far enough apart so that a tooth is always in between them to ensure there is never an opening from inlet to discharge which would increase slip.

## LUBRICATION

Journal bearings are the simplest type of bearing used in pumps. They are often called sleeve bearings because they are basically a sleeve that the shaft fits into. However, they are the most critical part of applying a pump to a particular application because they do not depend on rigid, mechanical parts for operation, but on a self-forming hydrodynamic film of liquid that separates the moving and stationary parts.

A journal bearing can operate in three ways.[8] First, it can operate with the shaft rubbing the bearing; this is called boundary lubrication. Second, the shaft may form a liquid film that completely separates it from the shaft; this is called hydrodynamic film operation. And finally, the third and most common mode is mixed film lubrication; this is where parts of the bearing and shaft are separated by a liquid film while other parts are in rubbing contact.

Boundary lubrication occurs at low shaft speeds or low fluid viscosities where the strength of the liquid film is insufficient to support the load on the bearing. Since the parts are rubbing under load, they will wear. Limits have been established that attempt to keep this wear down to acceptable levels. The limits are based on the amount of heat that can be removed from the bearing surfaces actually in contact, and on surface chemistry interactions between the shaft, bearing, and liquid. Boundary lubrication is often called "PV-lubrication" because the bearing load per unit area (P) and the relative velocity between the parts (V) are the factors. Typically, the following values are used to estimate load capabilities of various bearing materials:

| Material | PV-value (psi × ft/min) |
|----------|-------------------------|
| carbon   | 120,000 |
| bronze   | 60,000 |
| iron     | 30,000 |
| plastics | 3,000 |

Hydrodynamic operation is different from boundary lubrication in that the bearing material is not important, and the viscosity instead of the chemistry of the liquid is important. Bearing clearance also becomes a very important factor.

Roller bearings are often used on gear pumps where journal bearings will not work because of high loads, low speeds, or low viscosities. The bearing manufacturer's recommendations should be followed[8] in applying these bearings, except that most bearing manufacturers do not have information on bearing performance with lubricants below 150 SSU. In the absence of any better information, use the following rules of thumb:

- Liquid viscosity above 5 cSt — use full bearing catalog rating
- Liquid viscosity 5 to 2.5 cSt — use 75% of catalog rating
- Liquid viscosity 2.5 to 1 cSt — use 25% of catalog rating

(Note: cSt = viscosity in centistokes)

It may seem surprising, but the ends of the gears usually operate with hydrodynamic lubrication. The loads here are generally low, and the teeth form an excellent type of thrust bearing called a *step thrust bearing*. Wearplates can be used between the ends of the gears and the housing when the combination of gear and housing material could cause wear or galling problems. Galling is wear by adhesion of material from one part onto the other and is characteristic of stainless steel sliding against a similar metal under load. Wearplates can also be used when there is no hydrodynamic film on the ends of the gears. The wearplates will then be made from a material with good boundary lubrication characteristics, such as carbon or bronze.

Wear seldom occurs to the teeth of steel or cast iron gears unless there are abrasives present in the liquid, or when the gear teeth are heavily loaded and the material that they are made from does not have the strength necessary to resist deterioration of the tooth surface. Even with very low viscosity liquids, the loads on the gear teeth are normally low, and the contact is intermittent, so boundary lubrication is adequate. However, stainless steel gears generally exhibit severe wear[9] unless a full liquid film separates the teeth. Any particular pump at a certain speed will have a liquid viscosity below which the gears will begin to gall.

The choice of materials greatly affects the point where galling begins. Some materials are restricted to use only under full film lubrication conditions, while others can operate down well into the mixed film region without problems. Hardened martensitic stainless steels, particularly 440C, have good resistance to galling. 440C can be run against itself and is no more likely to gall than carbon steel. The other extreme is 18-8 austenitic stainless steels, like 304 or 316. These will gall when run against themselves or each other if there is any contact at all between the gears.

## USER COMMENTS

The following are some direct comments on various types of rotary pumps made by the users interviewed at the chemical plants, as they see it:

### EXTERNAL GEAR PUMPS

Workhorse of the industry, applied mostly for low capacity and high pressure. Range of viscosities is very wide: from lube oil to one million SSU. Application issues include close clearances between the gears and casing. Bearings are immersed in product and are therefore product lubricated, which makes them product sensitive. Both extremes of viscosity (very low and very high) may present lubrication problems. Seal chambers are restrictive, since gear pump manufacturers are only now beginning to come up with large sealing chambers — something centrifugal pump manufacturers have addressed 5 to 10 years ago. Due to these sealing chamber size restrictions, some standard mechanical seals may not fit in these pumps, without special design alterations. Gear pumps are often noisy in operation, and generate flow pulsations downstream.

Flow control of all positive displacement pumps is not as straightforward as for centrifugals. However, recent advances in variable frequency drives (VFD) make such method of flow control convenient, and relatively inexpensive. It also allows wiring the VFD signal to the operator control room, for remote control and monitoring.

Abrasive applications are an issue, just as for any other pump type. To combat abrasion, speed should be decreased, because wear is exponential with speed. Coatings include carburizing (low carbon steel), nitriding (alloy steel), or exotic materials (nickel alloys).

Cavitation is not a big issue for gear pumps, for many reasons. First of all, they typically operate at lower speeds. Second, the "positive filling" of the inlets makes them less sensitive to flow non-uniformity, separation, and backflow — which is a very serious factor for centrifugal pumps. Third, gear pumps typically pump oils, and similar fluids, the properties of which are such (e.g., latent heat of vaporization, characterizing the intensity of cavitation damage, is much lower) that even if cavitation does occur, the damage to the internals is less.[4]

The necessity to have the relief valve (internal in-built, or external) is a limitation. Without a relief valve, very high discharge pressures may result, damaging the pump or connecting piping, as well as cause safety issues. Therefore, a relief valve must always be present. The integral relief valve, which is a part of a pump, is not designed to regulate the flow, but only relieves occasional overpressure, for a limited time. If the relief valve opens, and the equipment is not shut-down relatively soon, or the problem is not corrected otherwise, fluid overheating could result in a matter of minutes or less. Attention to relief valve operation is very important.

## Internal Gear Pumps

Designed for low pressure at low speed. Drive gear is cantilever, vs. between-bearings arrangement of external gear pumps. A cantilever design requires a bigger shaft, to minimize deflections. This may shorten seal life.

These pumps are simple to assemble and repair, with little training, and, when applied appropriately, can do the job well and inexpensively.

The issues are similar to the external gear pumps — close clearances, with possibility of contact, wear or galling, if stainless construction (in which case wearplates made from carbon or bronze are required). In general, issues, benefits, and limitations are similar to those for external gear pumps.

## Sliding Vane Pumps

These are mainly applied for low viscosity liquids. However, pumping lube oils and gasoline is not uncommon. Vanes slide (i.e., adjust) to compensate for wear and are easy to replace. If not replaced on time, however, vanes can wear out to the point of breakage, causing catastrophic failures. Preventative maintenance, therefore, should include vane replacement, and should be done at regular, established intervals. Vanes are available in softer or harder materials — as required for various liquids, for mildly abrasive and corrosive applications. However, very little development work seems to have been done in vane material, and manufacturers should be encouraged to do so. These pumps can be extremely noisy at speeds over 300 RPM.

## Lobe Pumps

Similar to external gear pumps, but they require timing gears since the lobes are designed for no contact and do not transmit torque. Occasional contact can happen in reality, as differential pressure deflects the shaft, pushing gears to rub against the casing walls. Larger and stiffer shafts minimize such deflections, to prevent gear contact. Information about the rotors' deflection with differential pressure can be obtained from the manufacturer, or estimates can be done by the plant engineers. Lobe pumps are good for high displacements (flows) and low pressure applications, and are often used for blowers/vacuum applications, as well as food applications.

# 6 Multiple-Screw Pumps

Rotary screw pumps have existed for many years and are manufactured around the world. More demanding service requirements impose challenges on screw pump manufacturers to provide higher pressure or flow capability, better wear resistance, improved corrosion resistance, and lower leakage emissions. Better materials and more precise machining techniques as well as engineering innovation have led to improvements in all these areas.

The first screw pump built was probably an Archimedes design used to lift large volumes of water over small vertical distances; they are still manufactured and used for this service. Today's three-screw, high performance pump can deliver liquids to pressures above 4500 psi and flows to 3300 gpm with long-term reliability and excellent efficiency. Power levels to 1000 hp are available. Twin-screw pumps are available for flow rates to 18,000 gpm, pressures to 1450 psi, and can handle corrosive or easily stained materials, again at good efficiencies. Power ranges to 1500 hp operate on critical applications.

In multiple-screw pumps, each wrap of screw thread effectively forms a stage of pressure capability. High pressure pumps have 5 to 12 stages or wraps where low pressure pumps may have only 2 or 3 wraps. This staged pressure capability is illustrated in Figure 30. More wraps are incorporated in pumps designed for higher pressure service.

## THREE-SCREW PUMPS

Three screw pumps are the largest class of multiple-screw pumps in service today.[10] They are commonly used for machinery lubrication, hydraulic elevators, fuel oil transport and burner service, powering hydraulic machinery and in refinery processes for high temperature viscous products such as asphalt, vacuum tower bottoms, and residual fuel oils. Three-screw pumps, also are used extensively in crude oil pipeline service as well as the gathering, boosting, and loading of barges and ships. They are common in engine rooms on most of the world's commercial marine vessels and many combat ships. Subject to material selection limitations, three-screw pumps are also used for polymer pumping in the manufacture of synthetic fibers such as nylon and lycra. Designs are available in sealless configurations such as magnetic drives and canned arrangements (see Figure 31).

The magnetic drive screw pump is used extensively for pumping isocyanate, a plastic component which is an extremely difficult fluid to seal using conventional

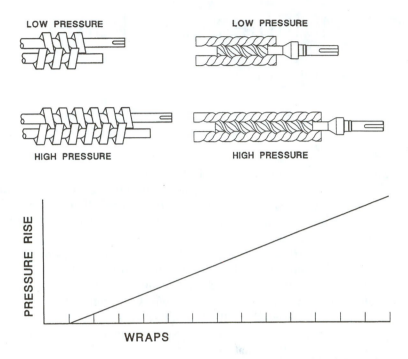

FIGURE 30  Staging effects of multiple-screw pumps. (Courtesy of IMO Pump.)

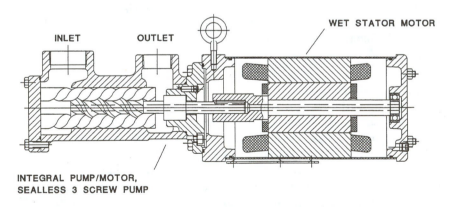

FIGURE 31  Three-screw canned motor pump. (Courtesy of IMO Pump.)

technology. Three-screw pumps are renowned for their low noise levels, high reliability, and long life. They are not, however, very low cost pumps.

### DESIGN AND OPERATION

Three-screw pumps are manufactured in two basic styles: single suction and double suction (see Figure 32).

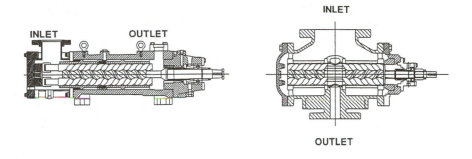

FIGURE 32 Single and double suctions three-screw pumps. (Courtesy of IMO Pump.)

The single suction design is used for low to medium flow rates and low to very high pressure. The double suction design is really two pumps parallel in one casing. They are used for medium to high flow rates at low to medium pressure.

Three-screw pumps generally have only one mechanical shaft seal and one, or perhaps two, bearings that locate the shaft axially. Internal hydraulic balance is such that axial and radial hydraulic forces are opposed and cancel each other. Bearing loads are thus very low. Another common characteristic of three-screw pumps is that all but the smallest low pressure designs incorporate replaceable liners in which the pumping screws rotate. This simplifies field repairs.

The center screw, called the power rotor, performs all the pumping. The meshed outside screws, called idler rotors, cause each liquid-holding chamber to be separated from the adjacent one except for running clearances. This effectively allows staging of the pump pressure to rise. Because the center screw is performing all the pumping work, the drive torque transferred to the idler rotors is only necessary to overcome the viscous drag of the cylindrical rotor spinning within its liner clearance. The theoretical flow rate of these pumps is a function of speed, screw set diameter, and the lead angle of the threads. Flow rate is a function of the cube of the center screw diameter. Slip, however, depends on clearances, differential pressure, and viscosity, and is only a square of the power rotor diameter. This results in larger pumps being inherently more efficient than smaller pumps, a fact that applies to most rotating machinery.

Speed is ultimately limited by the application's capability to deliver flow to the pump inlet at a sufficient pressure to avoid cavitation. This is true of all pumps. Three-screw pumps tend to be high speed pumps, not unlike centrifugal pumps. Two-pole (3600 RPM) and four-pole motors are most commonly used. When large flows, very high viscosities, or low available inlet pressures dictate, slower speed may be necessary. For example, some polymer services handle liquid at 250,000 SSU or more. Three-screw pumps on such service would typically be operated in the 50 to 150 RPM range. Gas turbine fuel injection service would more commonly be in the 3000 to 3600 RPM range since the fuels tend to be low in viscosity (1 to 20 centistokes), and the pump inlet is normally boosted to a positive pressure from a fuel treatment skid pump. The high speed operation is desirable when handling low viscosity liquids since the idler rotors generate a hydrodynamic liquid film in their load zones that resists radial hydraulic loads, very similar to hydrodynamic sleeve bearings found in turbomachinery.

In order to achieve the highest pressure capability from three-screw pumps, it is necessary to control the shape of the screws while under hydraulic load. This is best achieved by the use of five axis NC profile grinding which allows complete dimensional control and a high degree of repeatability. Opposed loading of the idler rotor outside diameters on the power rotor root diameter dictate that these surfaces be heat treated to withstand the cyclic stresses. Profile thread grinding allows the final screw contour to be produced while leaving the rotors quite hard, on the order of 58 RC (58 on the Rockwell C scale). This hard surface better resists abrasive wear from contaminants.

Because some three-screw pump applications range to pressures of 4500 psi, pumping element loading due to hydrostatic pressure can be quite high. Without hydraulic balance to counteract this loading in one or two planes, bearing loads would be excessive and operating life shortened.

Single-ended pumps use two similar, but different, techniques to accomplish axial hydraulic balance. The center screw, called a power rotor, incorporates a balancing piston at the discharge end of the screw thread (see Figure 33a). The area of the piston is made about equal to the area of the power rotor thread exposed to discharge pressure. Consequently, equal opposing forces produce zero net axial force due to discharge pressure, and place the power rotor in tension. The balance piston rotates within a close clearance stationary bushing, which may be hardened or hard coated to resist erosive wear. The drive shaft side of the piston is normally internally or externally ported to the pump inlet chamber. Balance leakage flow across this running clearance flushes the pump mechanical seal which remains at nominal pump inlet pressure. The two outer screws, called idler rotors, also have their discharge ends exposed to discharge pressure. Through various arrangements, discharge pressure is introduced into a hydrostatic pocket area at the inlet end of the idler rotors (see Figure 33b).

The effective area is just slightly less than the exposed discharge end area, resulting in approximately equal opposing axial forces on the idler rotors. The idler rotors are therefore in compression. Should any force cause the idler rotor to move toward discharge, a resulting loss of pressure, acting on the cup shoulder area or hydrostatic land area, tends to restore the idler rotor to its design running position. The upper view in Figure 33b shows a stationary thrust block (cross hatched) and a stationary, radially self-locating balance cup. Discharge pressure is brought into the cup via internal passages within the pump or rotor itself. The lower view shows a hydrostatic pocket machined into the end face of the idler rotor. It too is fed with discharge pressure. The gap shown is exaggerated and is actually very near zero.

For some contaminated liquid services, the hydrostatic end faces of the idler rotors are gas-nitride-hardened or manufactured from solid tungsten carbide, and shrink fitted to the inlet end of the idler rotors. When the cup design is used, the cup inside the diameter and shoulder area are normally gas–nitride hardened. Both techniques are used to resist wear due to the fine contaminants.

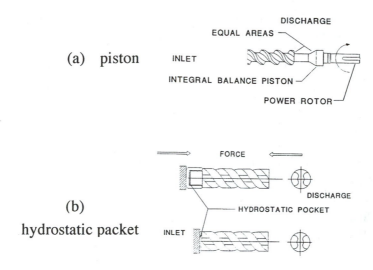

**FIGURE 33** Hydraulic thrust balancing. (Courtesy of IMO Pump.)

In a radial direction, three-screw pumps achieve power rotor hydraulic balance due to symmetry. Equal pressure acting in all directions within a stage or wrap results in no radial hydraulic forces, since there are no unbalanced areas. The power rotor will frequently have a ball bearing to limit end float for proper mechanical seal operation, but it is otherwise under negligible load. Idler rotor radial balance is accomplished through the generation of a hydrodynamic liquid film, and operates similar to a journal or sleeve bearing.

The eccentricity of the rotating idler rotor sweeps liquid into a converging clearance resulting in a pressurized liquid film. The film pressure acts on the idler rotor's outside diameters in a direction opposing the hydraulically generated radial load. Increasing viscosity causes more fluid to be dragged into the pressurized film, causing the film thickness and pressure supporting capability to increase. The idler rotors are supported in their respective housing bores on liquid films and have no other bearing support system. Within limits, if differential pressure increases, the idler rotor moves radially toward the surrounding housing bores. The resulting increase in eccentricity increases the film pressure and maintains radial balance of the idler rotors.

In three-screw pumps, inlet pressure above or below atmospheric pressure will produce an axial hydraulic force on the drive shaft. In most pump applications, pump inlet pressures are below, or slightly above, atmospheric pressure so the forces generated by this low pressure acting on a small area are negligible. However, if the application requires the pump to operate at an elevated inlet pressure, from a booster pump for example, then inlet pressure acting on the inlet end of the power rotor is only partly balanced by this same pressure acting on the shaft side of the balance piston. In effect, the area of the power rotor at the shaft seal diameter is an unbalanced area. This area multiplied by the inlet pressure is the resulting axial load, toward the shaft end. When specifying pumps, it is important to clearly state the maximum expected inlet pressure so the pump manufacturer can verify this loading. Several reliable methods are in use to carry this load, including antifriction bearing such as double extending the power rotor out the inlet end of the pump (which adds a second shaft seal) or sizing the balance piston to counterbalance this axial force.

As versatile as three-screw pumps are, they are not suitable for some applications. While many advances in materials engineering are taking place, the state of the art for three-screw pumps is such that very corrosion-resistant materials, such as high nickel steels, have too great a galling tendency. The rotors of three-screw pumps touch, and thus any materials that tend to gall are unsuitable. Unfortunately, this includes many corrosion-resistant materials. Viscosities too low to allow hydrodynamic film support generation are also application areas for which the three-screw pump is not optimal.

## TWO-SCREW PUMPS

Generally, two-screw (or twin-screw) pumps are more costly to produce than three-screw pumps and thus are not in as extensive use. They can, however, handle applications that are well beyond many other types of pumps, including three-screw designs. Twin-screw pumps are especially suited to very low available inlet pressure applications, and more so if the required flow rates are high. Services similar to three-screw pumps include: crude oil pipelining; refinery hot, viscous product processing, synthetic fiber processing; barge unloading; fuel oil burner and transfer; as well as unique applications such as: adhesive manufacture; nitrocellulose explosive processing; high water cut crude oil; multiphase (gas/oil mixtures) pumping; light oil flush of hot process pumping; cargo off-loading with ballast water as one of the fluids; and tank stripping service where air content can be high and paper pulp production needs to pump over about 10% solids.

### DESIGN AND OPERATIONS

The vast majority of twin-screw pumps are of the double suction design (see Figure 34). The opposed thread arrangement provides inherent axial hydraulic balance due to symmetry. The pumping screws do not touch each other and thus lend themselves well to manufacture from corrosion-resistant materials. The timing gears serve to both synchronize the screw mesh as well as to transmit half the total power input from the drive shaft to the driven shaft. Each shaft effectively handles half the flow, and thus half the power. Each end of each shaft has a support bearing for the unbalanced radial hydraulic loads. A few designs leave the bearings and timing gears operating in the liquid pumped. While this results in a significantly lower cost pump design, it defeats much of the value that twin-screw pumps bring to applications. The more common and better design keeps the timing gears and bearings external to the liquid pumped. They need not rely upon the lubricating qualities of the pumped liquid or its cleanliness. Four mechanical shaft seals keep these bearings and timing gears isolated and operating in a controlled environment.

Hydraulic radial forces on a two-screw pump rotor due to differential pressure are illustrated in Figure 35. The forces are uniform along the length of the pumping threads. These hydraulic forces cause deflection "y" for which running clearance

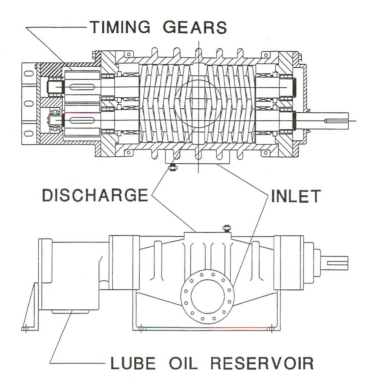

FIGURE 34 Double suction two-screw pump. (Courtesy of IMO Pump.)

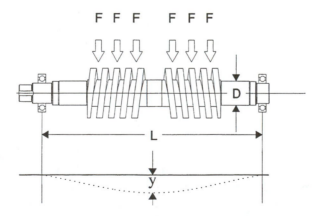

FIGURE 35 Radial forces in twin-screw pumps. (Courtesy of IMO Pump.)

must be provided in the surrounding pump body. Greater deflection requires larger clearances, resulting in more slip flow or volumetric inefficiency, so "y" must be kept to a minimum. Excessive deflection will cause damage to the surrounding body and/or contribute to rotating bend fatigue which will ultimately result in shaft breakage. The following is the general form of the deflection equation,

$$y = \frac{F \times L^3}{c \times E \times I} < \text{radial clearance} \qquad (39)$$

where "F" is the summation of the hydraulic forces, "L" is the bearing span, "c" is a constant, "E" is the shaft material modulus of elasticity, and "I" is the shaft moment of inertia. The shaft moment of inertia is a function of "$D^4$," where "D" is the effective shaft diameter. This equation is simplified and, in practice, must account for the varying shaft and screw diameters as they change along the length of the rotor. If screw "shells" are not integral with the shaft, that is, not made from a single piece of material, then material differences as well as attachment schemes must be factored into the deflection calculation. In any event, it is easy to see that the bearing span, "L," must be kept to a minimum to minimize deflection. The use of large diameter shafts and screw root sections helps to maintain minimum deflection.

Depending on the direction in which the threads are machined (left or right hand), and the direction of shaft rotation, the pump manufacturer can predetermine the deflection to be in either of the two radial directions: up or down, for a horizontal pump. These radial deflection loads are absorbed through externally lubricated anti-friction bearings. Higher differential pressure produces higher radial loads or forces. Smaller lead angles of the screw set reduce both these radial loads and the flow rate. Larger lead angles increase both flow rate and radial loading. Bearings are usually sized to provide 25,000 or more hours, "L10,"[8] bearing life at the maximum allowable radial loading and maximum design operating speed. Because of this pumpage-independent bearing system, two-screw pumps with external timing gears and bearings can handle high gas content as well as light oil flushes, water, etc.

Twin-screw pumps are manufactured from a broader range of materials, including 316 stainless steel. When extreme galling tendencies exist between adjacent running components, a slight increase in clearance is provided to minimize potential for contact. In addition, the stationary bores in which the screws rotate can be provided with a thick industrial hard chrome coating which further reduces the likelihood of galling as well as providing a very hard, durable surface for wear resistance. Such coatings do, however, require the capability of inside diameter grinding to achieve finished geometry within tolerances. For highly abrasive services, the outside diameter of the screws can be coated with various hard facings to better resist wear. Among these coatings are tungsten carbide, stellite, chrome oxide, alumina titanium dioxide, and others. Figure 36 shows a finished screw with a hard surfaced outside diameter. The high efficiency performance is a clear advantage over centrifugal pumps where liquid viscosity exceeds 100 SSU (20 centistokes). For pressures requiring two or more stages in a centrifugal pump, multiple-screw pumps frequently will be very competitive on a first cost basis as well. Operating liquid temperatures as high as 600°F have been achieved in twin-screw pumps for the ROSE® deasphalting process (see Figure 37). Timing gears and bearings are force cooled while the pump body is jacketed for a hot oil circulating system to bring the pump to process temperature in a gradual, controlled manner. Three-screw pumps have been applied to the same elevated temperature, more commonly on asphalt or vacuum tower bottoms services in refineries.

**FIGURE 36** Twin-screw thread shell with hard coated O.D. (Courtesy of IMO Pump.)

**FIGURE 37** Rose® process 600°F twin-screw pump. (Courtesy of IMO Pump.)

Medium and high viscosity operation are not the only regions where multiple-screw pumps bring advantages to the end user. Low viscosity, combined with high pressure and flows less than approximately 450 gpm, are excellent screw pump applications. Continuous, non-pulsating flow is required, for example, in high pressure atomizers for fuel combustion. Combustion gas turbines frequently burn distillate fuels, naphtha, and other low viscosity petroleum liquids that may reach

1 centistoke or less and require pumping pressures in the 950 to 1300 psi range. The combination of modest flow, low viscosity, and high pressure is a difficult service for all but reciprocating pumps. The pressure and flow pulsation from reciprocating pumps usually cannot be tolerated in fuel burning systems, especially combustion gas turbines.

Ongoing research and development efforts will further extend the capabilities of these machines allowing better performance over a broader range of applications. Multiple-screw pumps are uniquely suited to many of the applications described herein and offer long-term benefits to their users.

## USER COMMENTS

The following comments on various types of pumps were made by the chemical plant users interviewed:

> Multiple-screw pumps are the most technically complex and expensive of all rotary pumps. However, they have certain very important advantages. Applied to high pressures (2000 psi range) and high temperatures (designs are known to 800°F range), they can operate at a wide range of viscosities, from grease to rubber. Due to close clearance, these pumps have high efficiencies. Timing gears may, or may not, be required, depending on the design (two- vs. three-screw designs). Pinned rotors could be prone to failures but are a less expensive option. With external bearings and timing gears, four seals are required, and field work, for alterations and repair, is not trivial. Advantages of screw pumps are their quiet operation, low pulsations, and good NPSHR (net positive static suction head required) characteristics.

# 7 Progressing Cavity (Single-Screw) Pumps

The progressing cavity pump was invented as a supercharger for an airplane engine by Dr. René Moineau, following World War I. Due to the multiphase and high suction capability of the pump and its ability to convey large quantities of air, vapor, or gas in a fluid (which may also contain solids), this pump is not usually used in the same way or considered to be the same as other rotary positive displacement pumps. The progressing cavity pump was designed as a combination of a rotary pump and a piston pump. It represents an integration of the specific advantages of both types of pump constructions, such as high pumping flow rates, high pressure capabilities, minimal pulsation, valveless operation, and excellent pressure stability.[11] While the objective of this section is to introduce the pump user to the basic design principles, applications, operation, and maintenance of progressing cavity pumps, the focus on the theoretical will be limited and the emphasis will be on more practical, everyday issues. However, to get started we must understand how the pump operates, which requires some theory. (Additional information may be found in References 12 and 13).

## PRINCIPLE OF OPERATION

The progressing cavity pump is a helical pump belonging to the rotary, positive displacement pump family, as classified by the HI.[2] The pump consists of an internal thread stator with a double thread and an external thread rotor with a single thread. The meshing of the two forms a series of cavities 180° apart, which progress along the axis of the assembly as the rotor is rotated. As one cavity decreases in volume, the opposing cavity increases at exactly the same rate. Therefore, the sum of the two discharges is a constant volume. This results in pulsationless flow from the pump.

The cross section of the stator is two semicircles of diameter, "D," separated by a rectangle with sides "4e" and "D" (see Figure 38). The cross section of the rotor is a circle of diameter, "D," which is offset from the centerline by the eccentricity, "e." The pitch of thread of the stator is "Ps," and is twice the pitch of the rotor.

The dimensions of the cavity formed when the rotor and stator are meshed together is equal to the void of the cross section, filled with fluid:

$$A_{fluid} = 4eDP_s. \tag{40}$$

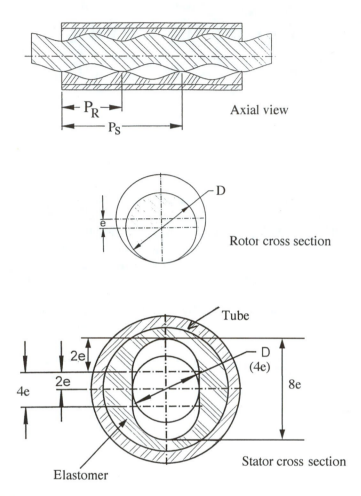

**FIGURE 38** Progressing cavity pump hydraulic sections (not shown to scale). (Courtesy of Roper Pump.)

This cross-sectional void multiplied by the stator pitch determines the cavity which is displaced upon each revolution of the rotor and can be expressed in various ways, one of the most popular being gallons per 100 revolutions.

The preceding description outlines the basic 1:2 design (i.e., the rotor having one thread or lobe, and the stator having two). Many variations exist on this such as 2:3, 3:4, 4:5, etc., designs. The only requirement is that the stator have one more thread or lobe than does the rotor. This book focuses only on the 1:2 design as this is the most popular and commercially available design for progressing cavity pumps available today. (Note: A good reference on multilobe designs is Reference 14.)

## PUMP PERFORMANCE CONSIDERATIONS

Being a positive displacement pump, a certain volume of fluid is discharged with each revolution of the rotor. Unlike centrifugal pumps, the pump does not develop pressure or a head but will attempt to deliver the same volume of fluid regardless of the pressure (resistance) that must be overcome in the discharge line. With a fluid such as water (1cP viscosity), and with zero discharge and suction pressures, the displacement, "q," or quantity of fluid delivered, of the pump is only dependent on the revolutions per minute.

As the pressure increases, a small amount of the fluid displaced slips back through the elements to the suction side. This slip is the fluid that leaks across the sealing lines of the cavities from the higher pressure discharge back to the lower pressure suction side of the elements. Slip is measured in units of flow such as gallons per minute (gpm). The amount of slip or leakage becomes greater as the discharge pressure increases. To minimize the amount of slip at high pressures, more cavities are added in series by lengthening the rotor and stator. This is called staging, analogous to multi-screw pumps, as discussed in earlier chapters. Increasing the number of stages decreases the amount of slip experienced by the pump by distributing the pressure differential over a greater number of pump stages. Staging is also used to obtain higher differential pressures for a given flow design. Past practice was to limit pressure per stage to 75 psi.

Catalog performance curves for progressing cavity pumps are usually for cold water and show the anticipated delivery of the pump versus differential pressures (discharge to suction) and speeds. Figure 39 shows a typical progressing cavity pump performance curve. For thicker, more viscous fluids, the slip is significantly reduced and can be approximated by dividing the slip on cold water by the slip index value shown in Figure 40. The slip index is the ratio of slip on a viscous fluid divided by the slip on water at 70°F.

To demonstrate the effects of viscosity on slip, assume the pump curve shown in Figure 39 is being used to pump a fluid with a viscosity of 10,000 cP at a differential pressure of 150 psi. From the curve in Figure 39, the slip on water at 150 psi is 9 gpm. This slip value is obtained by taking the flow at 0 psi and then subtracting the flow at 150 psi. (As the slip is basically independent of speed, this can be done at any speed as long as the speed is the same for both pressures in the formula.) Looking at the slip index curve in Figure 40, the slip index for 10,000 cP is about 6.2. Therefore, the slip on the pumped fluid will be approximately

$$\frac{\text{WaterSlip}}{\text{SlipIndex}} \text{ or } \frac{9}{6.2} \approx 1.5 \text{ gpm at this pressure.}$$

## POWER REQUIREMENTS

The power requirements for progressing cavity pumps are expressed differently by various manufacturers. Some manufacturers express the power requirements simply as horsepower required at a certain speed, and others express them as torque. Since the progressing cavity pump is considered a constant torque device when operating

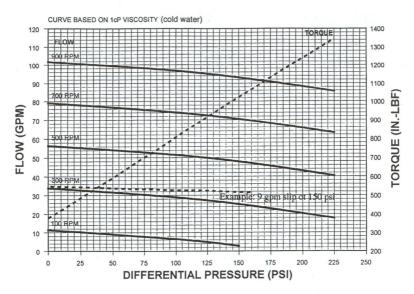

**FIGURE 39** PC-pump performance curve. (Courtesy of Roper Pump.)

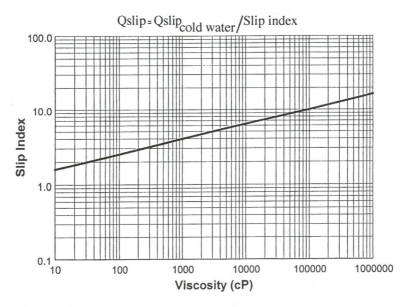

**FIGURE 40** Slip index curve. (Courtesy of Roper Pump.)

at a given differential pressure, using torque to determine power requirements is a simplified and accurate method of determining the power requirements and will allow better selection of drive components, especially hydraulic and electrical variable speed type drives.

The stator is usually made from an elastomeric material, although designs with a metallic (or rigid) stator for specialized applications are known. This elastomeric stator allows the pump elements to have a compression fit and also offers a good abrasion resistant surface for handling particles in suspension. This compression fit, however, does cause a resistance to turning (torque) which is dependent on the element geometry and is shown on the performance curve as the initial torque required at 0 psi. There is also a starting torque which must be overcome. As a rule of thumb, this value is roughly four times the initial torque. Horsepower can be derived from the torque value by the formula

$$Hp = \frac{TN}{63025},$$  (47)

where T is torque in inch pounds, and N is pump speed in RPM.

Progressing cavity pumps are rarely applied on fluids with a viscosity of 1 cP, and there is an added torque which is dependent on the viscosity of the fluid and the geometry of the elements. On slurries, the added torque required is dependent upon the particle size and concentration of solids. Most progressing cavity pump manufacturers include power adder information for viscosity and slurries with their standard technical literature. Typical power adder charts are reproduced below. These charts will, of course, vary with individual pump size and model.

**TABLE 1**
**Apparent Viscosity — Torque Additive (in × ebf) and Maximum Speed**

| cPs | 100 | 1000 | 2500 | 5000 | 10,000 | 50,000 | 100,000 | 150,000 | 200,000 |
|-----|-----|------|------|------|--------|--------|---------|---------|---------|
| T-add | 156 | 456 | 696 | 960 | 1335 | 2790 | 3840 | 4635 | 5325 |
| RPM | 900 | 900 | 900 | 600 | 320 | 80 | 40 | 30 | 25 |

**TABLE 2**
**Water Base Slurry Torque Additive (in × ebf)**

| Size | Fine | Medium | Coarse |
|------|------|--------|--------|
| % | 0.01" to 0.04" | 0.04" to 0.08" | 0.08" & Larger* |
| 10 | 175 | 219 | 325 |
| 30 | 526 | 657 | 976 |
| 50 | 877 | 1095 | 1627 |

*Note:* Maximum particle size 0.6 inch

## FLUID VELOCITY AND SHEAR RATE

With rotation of the rotor, fluid in the cavity moves in a spiral path along the centerline of the pump. The velocity of the fluid will be dependent on the speed of rotation as well as the distance from the centerline. Internal velocity information is available from manufacturers and is important for ensuring an adequate supply of fluid to the elements. The subsequent section on inlet conditions will discuss this aspect a bit more.

The shear rate of the pump can be determined from diameter, "D," eccentricity, "e," and the pitch of thread of the stator, "Ps," and is available from pump manufacturers for their particular models of pumps. Shear rate is defined as the difference in velocity between two layers of fluid, divided by the distance between the layers (measured in inverse seconds: $s^{-1}$ or $1/s$. This definition comes from the classical relationship between shear stress and velocity gradient, with dynamic viscosity being a proportionality constant.[15] In general, progressing cavity pumps have a very low shear rate. This is of particular importance when pumping shear sensitive fluids, such as latex and some polymers, that decompose if sheared too much during handling.

The shear rate of the pump can also be used to determine the apparent viscosity (the shear stress divided by the shear rate, measured in units of viscosity) of a non-Newtonian fluid inside the pump. This is useful in determining the power requirements of the pump, as explained in the next section.

## FLUID VISCOSITY

The dynamic viscosity ($\mu$) of a fluid is the property of a fluid that resists flow; it is the ratio of the shearing stress to the rate of shear. For fluids other than oil, the most common unit of measure for absolute (dynamic) viscosity is centipoise (cP). Sometimes the fluid viscosity is expressed in centistokes, which is the kinematic viscosity or the absolute viscosity divided by the specific gravity. When a fluid's viscosity is constant as the rate of shear is increased, it is said to be a Newtonian fluid. Most fluids that are handled by the progressing cavity pump do not obey this law and are said to be non-Newtonian. With a non-Newtonian fluid, the viscosity of the fluid changes as the rate of shear changes.

Some fluids will show a decrease in viscosity (thixotropic) as the rate of shear increases. The curves in Figure 41 show the viscosity in centipoise against the shear rate in inverse seconds. Often, when initially determining the requirements for a pump, the main emphasis is on operating point (i.e., where a pump would normally operate). Unfortunately, little attention is paid to other points along the performance curve, particularly the region of low flows, the assumption being that sizing of the equipment (pump and a driver) at operating conditions would ensure start-up. This may be a relatively safe approach for Newtonian and dilatent fluids, for which the starting torque is lower than the operating torque, and the driver (electric motor) would be able to start up the pump. For thixotropic fluids, however, such disregard may cause problems, since the shear stresses are higher near the start-up condition than at the operating regime. If the motor, in such a case, is sized for the operating conditions, it would not be able to start up the pump, in order to overcome the fluid resistance, having a higher viscosity at these conditions. Examples of thixotropic fluids are: adhesives, fruit juice concentrates, glues, animal oils, asphalts, lacquers, bentonite, lard, latex, cellulose compounds, waxes, syrups, fish oils, molasses, paints,

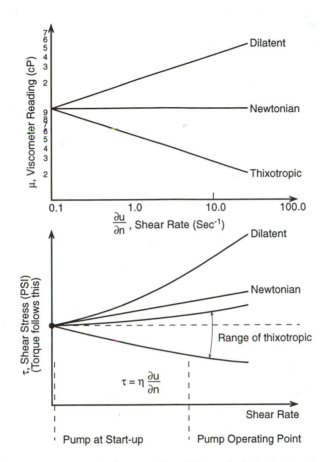

**FIGURE 41** Starting torque may be a problem, if viscosity behavior is unknown.

tar, rayon, printing inks, varnishes, resins, vegetable oils, and shortenings. An estimation of the apparent viscosity of the fluid can be made knowing several readings of the viscosity at known shear rates. The torque or power requirement for the pump can then be predicted at various speeds or shear rates.

Dilatent fluids are rather uncommon and are mostly high-concentrated slurries. A dilatent fluid increases in viscosity as the shear rate is increased. Again, the power requirements can be determined knowing the apparent viscosity.

There are certain fluids or materials which cannot be classified in the preceding categories and can be handled very nicely with the progressing cavity pump. These materials, such as filter cake, dewatered slurries or sludges, and paper stock are semi-dry and will not readily flow into the normal suction opening of the pump nor is it possible to obtain a viscosity measurement indicative of the thickness of the material. These applications are best handled by a pump where the standard suction housing is replaced with a flanged hopper and an auger is attached to the connecting rod to assist movement of the material into the pumping elements.

Usually the viscosity of a fluid changes when there are changes in fluid temperature. Lower temperatures usually increase a fluid's viscosity while higher temperatures reduce them. The temperature of the fluid must be considered in determining the viscosity of any fluid.

## INLET CONDITIONS

The key to obtaining good performance from a progressing cavity pump, as with all other positive displacement pumps, lies in a complete understanding and control of inlet conditions and the closely related parameters of speed and viscosity. To ensure quiet and efficient operation, it is necessary to completely fill the cavities with fluid as they open to the inlet. This becomes more difficult as viscosity, speed, or suction lift increases. Essentially, it can be said that if the fluid can be properly introduced into the pumping elements, the pump will perform satisfactorily. However, the problem lies in getting the fluid in. A pump does not pull or lift liquid into itself. Some external force must be present to initially push the fluid into the pumping elements. Normally, atmospheric pressure is the only force present, although there are some applications where a positive inlet pressure is available.

Naturally, the more viscous the fluid, the greater the resistance to flow, and, the slower the rate of filling the moving cavities in the inlet. Conversely, low viscosity fluids will flow quite rapidly and will quickly fill the cavities. It is obvious that if the cavities are moving too fast, the filling will be incomplete and a reduction in output will result. The rate of fluid flow into the cavities should always be greater than the rate of cavity travel, in order to obtain complete filling. Most progressing cavity pump manufacturers provide recommended maximum speeds for their pumps when handling viscous fluids that minimize the effects of incomplete filling.

The net positive suction head (NPSH) calculations are routinely used in centrifugal or high velocity pump applications. In many positive displacement pump applications where the pump velocities are usually low, the calculation of NPSH has little significance. There are applications, however, where this calculation becomes very important. These are suction lift, vacuum pot applications, and applications where the fluid vapor pressure is high.

The net positive suction head available (NPSHA) is the head available at the inlet of the pump and is the atmospheric pressure available minus the fluid vapor pressure, the lift, and suction line losses. The net positive static suction head required (NPSHR) is a function of the pump design and pump speed. The NPSHR for a specific pump is available from the pump manufacturer, usually included with standard performance information. Note, that in contrast to centrifugal pumps, the NPSHA definition for positive displacement pumps does not include (i.e., it disregards) a velocity head component, due typically to low velocities of PD pumps.

### SUCTION LIFT

The following is an example that illustrates NPSH calculations.

A particular progressing cavity pump operating at 900 RPM is required to lift 70°F water 10 ft vertically through a 3 in line. The vapor pressure of the fluid is .363 psia (.84 ft), and the suction line losses are .01 ft. The pump is physically

located at or near sea level. The pump manufacturer's literature indicates the pump
has an NPSHR of 6.9 ft. The calculations are as follows:

| | |
|---|---|
| Atmospheric pressure: | 33.90 ft |
| Lift: | −10.00 ft |
| Line losses: | −.01 ft |
| Vapor pressure: | −.84 ft |
| Total NPSHA: | 23.05 ft |
| Total NPSHR: | 6.90 ft |

From this example, there will be 16.15 ft of NPSH margin over the required amount,
and this will be acceptable.

## High Vapor Pressure Fluid

Another application where NPSH becomes important is when the fluid's vapor
pressure is high. Vapor pressure is a key property of fluids which must always be
recognized and considered. This is particularly true of volatile petroleum products
such as gasoline, which has a very high vapor pressure. The vapor pressure of a
fluid is the absolute pressure at which the fluid will change to vapor (boil) at a
given temperature. A common example is the vapor pressure of water at 212°F,
which is 14.7 psia. For petroleum products, the Reid vapor pressure (absolute) is
usually the only information available. This is vapor pressure as determined by
the ASTM D323 procedure. Vapor pressure tables for water can usually be found
in most hydraulic books.[16] Vapor pressure estimates for other fluids can be made
if the boiling point is known (i.e., a fluid will boil at atmospheric conditions when
its vapor pressure reaches 14.7 psia). The absolute pressure must not be allowed
to drop below the vapor pressure of the fluid. This will prevent boiling which will
cause cavitation.

   In the previous example, 70°F water with a vapor pressure of .363 psia was
used. If the temperature of the water was 190°F, the vapor pressure would be 9.34
psia or 21.6 ft, which would exceed the 16.15 ft margin between NPSHA and
NPSHR, and the fluid would vaporize or boil. The calculations would now be as
follows:

| | |
|---|---|
| Atmospheric pressure: | 33.90 ft |
| Lift: | −10.00 ft |
| Line losses: | −.01 ft |
| Vapor pressure: | −21.60 ft |
| Total NPSHA: | 2.29 ft |
| Total NPSHR: | 6.90 ft |

In this example, there is not enough NPSHA to operate the pump. To overcome
this situation, the amount of lift would have to be shortened or, possibly, a different
sized pump, operating at a reduced speed, could be used.

## VACUUM POT INSTALLATIONS

In a vacuum pot application, the fluid is in a vessel that is under a high or partial vacuum. This will affect the NPSHA to the pump. The following information is another example that illustrates NPSH calculations.

A particular progressing cavity pump is pumping water out of a vessel that is under 20 in of mercury vacuum, which corresponds to 11.20 ft of absolute pressure. There is 10 ft of 3 in horizontal line connecting the suction of the pump to the vessel, and the water is 70°F. As in the suction lift example, the vapor pressure of the fluid is .363 psia (.84 ft), and the suction line losses are .01 ft. From the pump manufacturer's literature, the pump NPSHR is 6.9 ft. The corresponding calculations are as follows:

| | |
|---|---|
| Vessel pressure: | 11.20 ft |
| Lift: | – 0.00 ft |
| Line losses: | –.01 ft |
| Vapor pressure: | –.84 ft |
| Total NPSHA: | 10.35 ft |
| Total NPSHR: | 6.90 ft |

From this example, there will be 3.45 feet of head over the required amount, and this would be acceptable.

On vacuum pot and suction lift applications, it is necessary to fill or partially fill the suction housing and lines with fluid to provide a lubricant for the elements during lift or until the pumping fluid reaches the elements. Since there are more sealing points on the side that is normally considered the suction side of a progressing cavity pump, better operation can be achieved if the pump is operated in reverse. By doing so, the normal discharge port is used as the suction port. This would now put the packing or seal[17] under pressure from the discharge, and caution should be used not to overpressure the suction housing. Progressing cavity pump manufacturers can provide helpful information on operating their pumps in reverse and should definitely be consulted if high discharge pressures are anticipated.

To allow pump manufacturers to offer the most economical selection and also assure a quiet installation, accurate inlet conditions should be clearly stated. Specifying a higher suction lift than actually exists may result in the selection of a pump at a lower speed than necessary. This means not only a larger and more expensive pump, but also a costlier driver. If the suction lift is higher than stated, the outcome could be a very noisy pump installation coupled with higher maintenance costs.

## GUIDANCE FOR PROPER SELECTION AND INSTALLATION

### ABRASION

The progressing cavity pump is one of the best pumps available for handling abrasive slurries; however, there are some considerations in pump size that need to be made in order to achieve maximum performance. It is necessary to minimize the slip and internal velocities to achieve good results. Most progressing cavity pump manufacturers limit these items by publishing reduced maximum speeds and pressures for pumps handling abrasive materials.

Determining the degree of abrasion is largely judgmental; however, the make-up of the particles will offer some clues as to how it is to be classified. A closer look at what causes abrasion may be helpful in determining its classification. The components of abrasion are the particle, the carrier fluid, and the velocity.

## Particles

*Size* — Wear increases with particle size.

*Hardness* — Wear increases rapidly with particle hardness when it exceeds the rotor surface hardness.

*Concentration* — The higher the concentration, the more rapid the wear.

*Density* — Heavier particles will not pass through the pump as easily as lighter ones.

Relating the material hardness to some common materials on a 1 to 15 scale, with 15 being the hardest, the following list can be used as a guide:

| | |
|---|---|
| Talc slurry | 1 |
| Sodium sulfate | 2 |
| Drilling mud | 3 |
| Kaolin clay | 4 |
| Lime slurry | 5 |
| Toothpaste, potters' glaze | 6 |
| Gypsum | 7 |
| Fly ash | 8 |
| Fine sand slurry | 9 |
| Grout, plaster | 10 |
| Titanium dioxide | 11 |
| Ceramic slurry | 12 |
| Lapping compound | 13 |
| Emery dust slurry | 14 |
| Carborundum slurry | 15 |

## Carrier Fluids

*Corrosivity* — Surfaces attacked by corrosion will set up a corrosion-erosion effect.

*Viscosity* — A high viscosity fluid will tend to keep particles in suspension and not be as abrasive.

*Velocity* — Low fluid velocity, or pump speed, will minimize abrasive effects. (For a heavy abrasive fluid, it is recommended to keep the average velocity in the elements between 3 to 5 ft/sec. A medium abrasive fluid should be limited to 6 to 10 ft/sec, and a light abrasive should be limited to 10 to 15 ft/sec. These velocity limits are usually listed in catalogs as pump rpm limits for various size pumps in conjunction with the abrasive nature of the fluid.)

As mentioned earlier, determining the degree of abrasion is largely judgmental: however, the examples below will provide a simple rule of thumb guide to abrasive classification:

*No Abrasives* — Clear water, gasoline, fuel oil, lubricating oil.
*Light Abrasives* — Dirty water containing silt or small amounts of sand or earth.
*Medium Abrasives* — Clay slurries, potters glazes, porcelain enamel, sludge.
*Heavy Abrasives* — Plaster, grout, emery dust, mill scale, lapping compounds, roof gypsum.

As with the shear rate, the maximum particle size that can be handled by a progressing cavity pump can be determined from the element design and is available from pump manufacturers for their particular models of pumps. Depending on pump size, the maximum particle size that can be successfully handled without pump damage ranges up to about 2 in.

## TEMPERATURE EFFECTS AND LIMITS

The fluid temperature will affect the pump performance in two different ways. First, since the stator is an elastomeric material, the thermal expansion is roughly ten times greater than that of metal rotor which is usually steel or stainless steel. This will cause a tighter fit for the elements, and higher starting and running torques. When the temperature reaches a certain limit it is then advisable to use an undersized rotor which compensates for the difference in size. Second, the life of the elastomer is greatly affected by heat. Table 3 shows limits from one manufacturer for elastomers which are being worked such as in a stator, and will differ from other published information on elastomers used in a static state such as o-rings and gaskets. There-fore, when a stator is being applied at less than its maximum pressure rating, the operating limit can be exceeded slightly but cannot exceed the static rating.

## MOUNTING AND VIBRATION

The progressing cavity pump is inherently an unbalanced machine due to the eccen-tric rotation of the rotor. The vibration which occurs is dependent on the size of the element, the offset, and the speed of rotation. For this reason, the speed of the pump is limited with typical limiting speeds reaching 1200 RPM for small pumps (.02 gal/rev and lower), decreasing to limits of 300 RPM for large pumps (3 gal/rev and larger).

**TABLE 3**
**Stator Temperature Ratings**

| | Temperature Rating (°F) | |
| Material | Stator Rating | Maximum (Static) Rating |
|---|---|---|
| Nitrile | 180 | 250 |
| Natural rubber | 185 | 225 |
| EPDM | 300 | 350 |
| Fluoroelastomer | 300 | 400 |

The magnitude of the induced vibration is of a low frequency and a relatively high amplitude. This will not produce offensive noise; however, it should be a factor in mounting. It is generally recommended that progressing cavity pumps be mounted on structural steel baseplates securely lagged down to concrete foundations. These baseplates should be provided with means to grout them into place on the foundation for rigidity and dampening. In all cases, the manufacturer's instructions and recommendations on mounting should be consulted to ensure an adequate installation.

## THE DRIVE FRAME

The progressing cavity elements are normally adapted to a drive end which will provide an acceptable life span when properly applied and maintained. As with ANSI centrifugal pumps,[18] the drive shaft is supported in a separate bearing housing (usually cast iron), by antifriction bearings (either deep groove ball or tapered roller bearings). These bearings are normally sealed from contamination by conventional lip type seals.[19] The bearings in a progressing cavity pump are almost always lubricated by grease. Due to the relatively low speed of progressing cavity pump applications, this is quite acceptable.

Because of the rotor's eccentric motion, there must be some means of transferring the torque from the concentric rotating shaft to the rotor. The usual means of achieving this is by using a connecting rod and universal joints. This combination allows the rotor to move freely through the eccentric circle and carries the torque and thrust created by the differential pressure on the elements. Several types of universal joints are used, with the two most popular being a simple pin and rod joint, and the other a more elaborate crowned gear joint. Some manufacturers even offer a flexible shaft, eliminating the universal joints entirely.

The pin and rod pumps are considered the more traditional type pumps (see Figure 42). Pins are installed at both ends of the connecting rod to secure it to the shaft and rotor. In operation, the connecting rod rocks on the pins from the eccentric motion of the rotor. Rubber o-rings or some other seal (often proprietary) keeps pumpage from the joint and retains lubricant. The pins and connecting rod are made from heat-treated alloy steels to provide a long service life. When a stainless material is required for corrosion resistance in the pumping application, the pins and rod cannot usually be hardened. Most manufacturers can increase the drive frame size

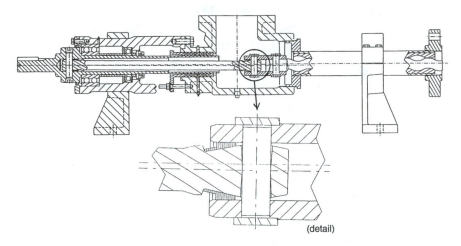

(detail)

**FIGURE 42** Typical pin joint pump. (Courtesy of Roper Pump.)

of pumps requiring these corrosion-resistant materials in order to provide an accept-able service life. If differential pressures are relatively low (less than 75 psi), then this increase in frame size may not be necessary.

Unfortunately, the rocking between the pin and rod is a wearing action that eventually renders the pin unusable. To overcome this and prolong service life, progressing cavity pump manufacturers also employ a crowned gear type joint which has approximately five times the load carrying capacity of the conventional rod and pins (see Figure 43). The gear joint is essentially a modified ball and socket joint, with gear teeth on the ball mating with a similar tooth profile in the socket. These joints are usually designed for the larger size pumps ($\approx 0.1$ gal/rev and larger) where the initial expense is outweighed by the maintenance and replacement costs of the rod and pin type.

## PROGRESSING CAVITY PUMP APPLICATIONS

The number of applications for progressing cavity pumps are as many and as varied as there are progressing cavity pumps manufactured. As mentioned earlier, progress-ing cavity pumps have many applications in industrial processes, construction, waste-water treatment, oil field service, pulp and paper processes, and food processing, just to name a few. Flow rates available range from fractions of gal/min, to around 2000 gpm, with differential pressures reaching upward to 2000 psi.

As with every pump type, there are certain advantages and disadvantages char-acteristic of the progressing cavity design. These should be recognized in selecting the best pump for a particular application.[20]

*Advantages:*

- Wide range of flows and pressures.
- Wide range of liquids and viscosities.

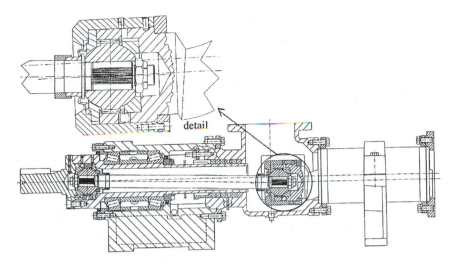

**FIGURE 43** Typical gear joint pump. (Courtesy of Roper Pump.)

- Low internal velocities.
- Self-priming, with good suction characteristics.
- High tolerance for entrained air and gases.
- Minimum churning or foaming.
- Pulsation-free flow and quiet operation.
- Rugged design — easy to install and maintain.
- High tolerance to contamination and abrasion.

*Limitations:*

- High pressure capability requires long length of pumping elements.
- Fluid incompatibility with elastomers can cause problems.
- Not suited to high speed operation — requires gear reducer or belt reduction.
- Cannot run dry.
- Temperature limitations.

In oil field service, progressing cavity pumps are used in many applications in the production of oil, from polymer injection to crude oil transfer. China's Daqing Petroleum Administrative Bureau installed a battery of large progressing cavity pumps for polymer transfer in the Daqing oil field, located about 750 miles northeast of Beijing. Each of these pumps is used to transfer 250 gpm of 3000 to 5000 cP polymer at discharge pressures of 300 psi. The pumps selected were of the 1.15 gal/rev size and operate at 250 RPM with 130 hp drivers.

Another interesting oil field application is the progressing cavity "down hole" pump. This is a progressing cavity pump suspended vertically below the surface into an oil well to pump fluid to the surface.[21] Driven by a surface-mounted driver, the pumps are suspended to depths of up to 4000 ft. Able to pump the entrained gases and sands, progressing cavity down hole pumps are an economical alternative to conventional rod pumps in many wells with difficult-to-pump fluids. They are proving to be particularly well suited to the heavy crudes in the oil fields of Canada.

Progressing cavity pumps are also useful in mining operations worldwide. In Australia, progressing cavity pumps are used in the gold mines for dewatering service in the mines. Located about 700 ft underground, these pumps deliver 560 gpm at a discharge pressure of 330 psi to keep the mines from flooding. The water contains about 5% earth solids, with an average particle size of .040 in. The units selected for this service were in the 1.75 gal/rev size and operate at 340 RPM, using 175 hp drivers.

The construction industry is a large user of progressing cavity pumps. In Salt Lake City, Utah, spackling (wall joint compound) with an apparent viscosity of 30,000 cP is pumped at $1^{1}/_{2}$ to 2 gpm against discharge pressures of 80 psi. The units selected for this service were in the 0.02 gal/rev size and operate at 140 RPM using 3 hp drivers.

Another construction application involves grout injection. In Southern California, portable units are made to inject grout under bridges to shore them up without closing them to traffic. These pumps deliver 35 gpm against 250 to 350 psi discharge pressures. The units selected for this application were in the .22 to .28 gal/rev size and operate at 175 RPM, using 25 hp drivers. Many manufacturers of grout equipment use progressing cavity pumps; this represents a large segment of their usage.

Wastewater treatment is another area where progressing cavity pumps are widely used. Their use ranges from scum and sludge handling to polymer injection. In one Illinois correctional facility, units are installed in the 0.28 gal/rev size operating at 400 RPM, using 10 hp drivers. These units deliver 100 gpm against a discharge pressure of 30 psi.

The pulp and paper industry uses a wide variety of progressing cavity pumps in the making of paper. Coating kitchen applications is a prime example of this. Both Ohio and Mississippi paper plants use progressing cavity pumps to pump paper coatings in their coating of kitchens. The units are of the 1.15 gal/rev size and operate at 235 RPM, using 75 hp drivers. These units deliver 200 gpm against a 170 psi discharge pressure.

In food processing, many chicken processing plants use progressing cavity pumps to pump chicken parts (heads, feet, eviscera, etc.) as waste. One Arkansas plant uses several units sized for 0.44 gal/rev operating at 200 RPM, to deliver 80 gpm against a 100 psi discharge pressure. These units are driven by a 20 hp driver. These pumps use a specially enlarged suction housing and auger feed to ensure that the material reaches the pumping elements.

Progressing cavity pumps are not only limited to handling waste in food processing plants. They are also manufactured in food grade designs to pump foods and pharmaceuticals. Pumps used for these services offer quick disassembly for cleaning and are made from food grade materials. Some manufacturers offer designs to meet 3-A sanitary standards. Applications for these units range from pumping toothpaste to maraschino cherries, the latter with less than a 1% damage rate.

## TROUBLESHOOTING

The progressing cavity pump is usually forgiving in most installations, but there are times when troubles occur. Accurate identification of the problem is usually 99% of the cure. The following guides will assist in identifying the problem and offer some suggestions for rectifying the situation. The guides can assist in tracking down potential troubles, and most of them are useful for all rotary positive displacement pumps, not only progressing cavity pumps.

### *No Liquid Delivered:*

- Pump rotating in wrong direction.
- Inlet lift too high; check this with gage at pump inlet.
- Clogged inlet line.
- Inlet pipe not submerged.
- Air leaks in inlet line.
- Faulty pressure relief device in system.
- Pump worn.

### *Rapid Wear:*

- Excessive discharge pressure.
- Pump runs dry.
- Incompatibility of liquid and pump materials.
- Pipe strain on pump.
- Speed too high for abrasives present in liquid.

### *Excessive Noise:*

- Starved pump.
- Air leaks in inlet line.
- Air or gases in liquid.
- Pump speed too high.
- Improper mounting; check alignment thoroughly.

### *Pump Takes Too Much Power:*

- Speed too high.
- Liquid more viscous than previously anticipated.
- Operating pressure higher than specified; check this with gage at pump discharge.
- Discharge line obstructed.
- Mechanical defect such as bent shaft.
- Packing too tight.
- Pipe strain on pump.

- Incompatibility of liquid and pump material causing stator swell.
- Pressure relief device in system not operating properly.

*Rotors:*

Other than normal wear on a rotor, there is little that can cause a problem. Occasionally, however, a problem will arise with the chrome plate lifting. This is usually caused by corrosion of the base metal of the rotor. In these situations, a stainless steel rotor is usually the solution. On fluids of pH 3 or less, the fluid will remove the chrome and a non-plated rotor should be used.

*Stators:*

Three modes of stator failure are bond failure, chemical attack, and high heat:

- *Bond Failure* — In the stator manufacturing process, a special curing type adhesive is applied to a grit blasted tube prior to injection of the elastomer. The bond to the tube is stronger than the tensile strength of the elastomer and is sufficient to hold the rubber in place during operation. On fluids with a pH of 10 or greater, there exists the potential of chemical attack on the bond, and it will eventually break loose from the inside of the tube. Very high heat (as from an acetylene torch) or cryogenic temperatures will also break the bond.
- *Chemical Attack* — The effect of chemical attack is usually a swelling and softening of the elastomer. The pump operates well at first, and then gradually increases in power requirements as swelling occurs, until the elastomer fails. Analysis of this problem should be made immediately, because after "drying out," the elastomer will return to its original state and appear fine. When there is doubt about material compatibility, small test slugs of elastomer can be immersed in the fluid to study the effects before the pump is purchased. These test slugs are available from many progressing cavity pump manufacturers.
- *High Heat* — The enemy of rubber is heat, and it will have a different effect on different materials. Nitrile and fluoroelastomer will harden to a hardness similar to steel and will have a glossy and crazed appearance. EPDM and natural rubber will have a tendency to get gummy and have a melted appearance. The source of excessive heat can come from several sources. The fluid temperature can go over the allowable limit or the heat can be internally generated. This internal heat can come from overrating the pump, pumping against a closed discharge, or running dry. In the first two cases, the stator usually will show signs of coming apart in chunks.

Progressing cavity pumps are widely used in handling many difficult-to-pump materials. In many cases, a progressing cavity pump is the only pump that can suitably handle these materials. Properly understanding how this pump works and how to specify the operating characteristics can help the user to assist progressing cavity pump manufacturers in selecting the most economical unit to solve many difficult pumping problems. When properly selected, progressing cavity pumps are quiet in operation and forgiving of many operational problems that arise in the system. Initially an oddity, progressing cavity pumps now are gaining a foothold in the mainstream of pumping applications.

## USER COMMENTS

The following comments were made by pump users interviewed at chemical plants.

These pumps are excellent for abrasive services, as well as to pump suspended solids. They can pump sewage, filter coat application, and other tough pumpages. Low shear and low turbulence make them a good choice for the shear-sensitive fluids. Viscosity ranges to extremes — to 5 million SSU. Bearings are mounted externally, outside of pumpage, eliminating the dependence on lubricating properties of the pumpage.

Elastomer (liner) compatibility with pumped fluids is important, and dry running can be catastrophic, and fast. Seal design, although better than for other rotary pumps, is still space restrictive, and would benefit by the manufacturers' attention to this issue. Volumetric efficiency is high, due to tight clearances and low slip. However, friction between the rotor and stator can negate the benefit of tight slip, due to friction losses, resulting in lower mechanical (overall) efficiency. These rubbing concerns limit the RPM to around 300 to 500 RPM, or lower for abrasive applications, and require a gear reducer or a VFD. These pumps require a considerable footprint area, which could be an issue if space is limited. However, if installed at an open yard area, this is not a problem.

# 8 Metering Pumps

(Note: Due to the specialized application of these pumps, and in an effort to be complete, this chapter extends beyond just rotary types and into the area of other types of positive displacement pumps, such as piston/plunger, diaphragm, peristaltic, etc.)

## DEFINITION

The metering pump is a controlled volume, positive displacement pump, which will accurately transfer a predetermined volume of fluid (liquid, gas, or slurry) at a specified amount or rate into a process or system. Metering pumps are capable of both continuous flow metering and dispensing.

## FEATURES

- Positive displacement imparts energy to the liquid in a pulsating fashion rather than continuously.
- Flow rates are predetermined accurately and are repeatable within ± 1% (*within the specified range and turn-down ratio*).
- Adjustable volume control is typically inherent in the design.
- Traditionally they were reciprocating (back & forth action) designs, although rotary configurations (gear and peristaltic) are increasingly applied to metering applications.
- Ideally, a metering pump should be capable of handling a wide range of liquids, including those that are toxic, corrosive, dangerous, volatile, and abrasive.
- Should be capable of generating sufficient pressure to permit injection of liquids into processes.

## WHERE ARE METERING PUMPS USED?

Metering pumps are utilized in virtually every segment of industry to inject, transfer, dispense, or proportion fluids. The chemical process industries have by far the greatest diversity of applications. However, in addition to a broad range of industrial applications, metering pumps and dispensers are used extensively in laboratory, analytical instrumentation, and automated medical diagnostic equipment:

- *Industrial* — Chemical, petroleum, electronics, water and wastewater treatment, food processing, agriculture, metal finishing, aerospace, automotive, mining
- *Analytical Instrumentation* — Chemical analyzers, environmental monitors
- *Medical* — Diagnostic systems, dialysis, disposable component assembly

## TYPES OF METERING PUMPS

Although a wide variety of both pumping and non-pumping designs are used to meter or proportion fluids, not all are considered metering pumps.[22] Pump designs, such as centrifugal, can be used to transfer fluids but would not be considered a metering pump. Venturi tubes or proportioning valves are examples of non-pump metering designs. Accuracy of these systems typically vary between 5 and 10%, which for certain applications is acceptable. What these systems lack the accuracy and repeatability inherent in positive displacement pump designs which transfer fluid in units of controlled volume.

The following designs of pumps are generally accepted as metering pumps. Although their designs differ, they are all positive displacement and are designed for precision fluid transfer — repeatable accuracy.

- *Packed Plunger* — reciprocating plunger or piston moves fluid through inlet and outlet check valves
- *Diaphragm* — uses a flexible diaphragm to move fluid through inlet and outlet check valves
- *Gear Pumps* — use the spacing between gear sets to move fluid
- *Peristaltic Tubing Pumps* — use rollers to move fluid through flexible tubing
- *Rotating/Reciprocating: Valveless Pump* — uses only one moving part, a rotating reciprocating piston, both to move fluid and to accomplish valving functions

## COMPONENTS OF METERING PUMPS

### THE POWER END OR DRIVE

The power end or drive is the source of power for the pump:

- *Electric Motors* — AC, DC, stepper motors; can be fixed or variable speed.
- *Pneumatic* — Pneumatic cylinder piston is controlled by valves or pneumatic rotary drive.
- *Electromagnetic* — Reciprocating linear motion is driven by solenoids.
- *Mechanical* — Pedestal pumps are driven by chains, belts, and gears.

## POWER CONVERSION OR DRIVE MECHANISMS

Power conversion or drive mechanisms convert power supplied from the drive to the type of motion required for the pump head. Reciprocating and rotary designs each use different methods to accomplish the same goal.

### Reciprocating Design

A reciprocating design in a metering pump converts the drive power, usually rotary action of the drive to linear reciprocating movement. This drive mechanism provides adjustable displacement volume by mechanically adjusting the stroke length. Flow adjustment is accomplished both mechanically and/or by drive speed (within pump head flow range).

- *Amplitude Modulation Mechanisms* — Plunger, piston and diaphragm pumps
- *2-Dimensional Slider Crank* — The radius of eccentricity of the pivot arm, crank, or cam is in the same plane as the motion of the connecting rod. Adjustment of the stroke is accomplished by varying the crank radius (see Figure 44).

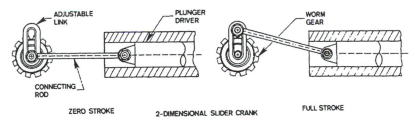

**FIGURE 44** 2-Dimensional slider crank.

- *3-Dimensional (Polar) Slider Crank* — The rotating crank radius is fixed. The plane of rotation of the drive mechanism is 90° to the motion of the connecting rod at zero stoke, and changes proportionally as the drive angle increases or decreases from the 90° position (see Figure 45).

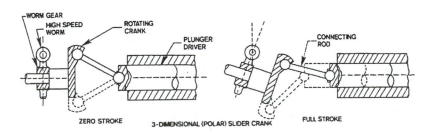

**FIGURE 45** 3-Dimensional slider crank.

- *Lost Motion Drive Mechanisms* — They are mechanisms which limit stroke (see Figure 46).
- *Mechanical (Eccentric Cam Drive)* — Provides for changes in the flow rate through variation of the plunger return position.
- *Hydraulic Bypass* — Permits a portion of hydraulic fluid to escape through a bypass valve which returns this unused fluid to a reservoir within the pump.

LOST MOTION DRIVES

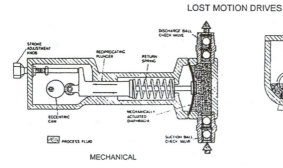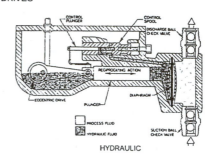

**FIGURE 46** Lost motion drives.

## Rotary Design

Direct, geared, or magnetically coupled, rotary drive mechanisms are used to couple gear or peristaltic pump heads to drives. Flow adjustment is accomplished by changing the pump head displacement components (gear set, tubing size), or through variable speed drives.

- *Flow Rate = Volume/Stroke × Stroke Rate* — For pumps with adjustable displacement, flow rate can be varied without varying drive speed (RPM). For pumps which do not have adjustable volumes, flow is varied by adjusting the drive speed. Gear and peristaltic pumps are examples of pumps which vary flow directly by varying drive speed.
- *Flow Range* — Typically fixed for a specific size displacement element. The displacement element could be a diaphragm, bellows, piston, gear set, peristaltic tubing, etc.
- *Turn-down Ratio* — The ratio of the maximum flow to the minimum flow value, where, at the minimum, accuracy is maintained.

### PUMP MECHANISM (PUMP HEAD MODULE)

A pump mechanism is a transfer mechanism which is in contact with the fluid and gives each specific pump technology its unique characteristic.

## Packed Plunger Pumps and Piston Pumps

In these pumps, a reciprocating piston (plunger) moves fluid through a chamber by creating alternate suction and pressure conditions. One-way check valves on the inlet and outlet ports of the pump operate 180° out of phase in order to control filling of the displacement chamber during suction and to prevent backflow during the discharge stroke (see Figure 47).

*Primary Advantages* — high pressures, accuracy, self-priming
*Limitations* — packings require maintenance, check valves, moving parts in
   contact with fluid

### PACKED PLUNGER PUMP

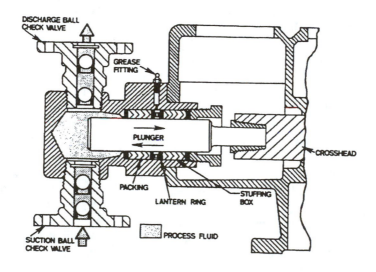

### PISTON PACKED PUMP

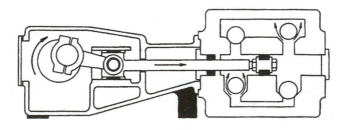

**FIGURE 47** Packed plunger and piston packed pumps.

## Diaphragm Pumps

In these pumps, a flexible disc, tube, or bellows is actuated by a connecting rod, eccentric cam, or a hydraulically coupled plunger. The fluid chamber is designed so that the diaphragm is one of the fluid containment components. As it reciprocates, the diaphragm changes the volume of the fluid chamber. This in turn creates alternate suction and pressures conditions in the chamber, which with the aid of check valves, draws fluid into the inlet port during suction, and moves fluid out of the discharge port when the pressure increases.

- *Mechanical Diaphragm* — The power side of the pump is identical to piston and plunger pumps. However, in place of a piston rod or plunger, the mechanical diaphragm pump uses a connecting rod fastened to the center of the diaphragm. The configuration of the diaphragm can take on many forms, but the most popular designs are the flat disc, the convoluted disc, and the bellows.
- *Hydraulic Diaphragm Pump* — Hydraulically balanced, it is a hybrid of the piston and mechanical diaphragm pumps described previously. The power (or drive) end, as well as capacity control, are the same. It is also similar to the plunger pump in that a plunger or piston reciprocates within a precision cylinder (also referred to as a chamber or liner). It is similar to the mechanical diaphragm in that a diaphragm contacts the process fluid rather than the plunger. At a set stoke length, the plunger displaces a precise volume of hydraulic oil, which in turn exerts pressure on a diaphragm. The hydraulic oil moves the diaphragm forward and backward, causing a displacement that expels the process fluid through the discharge check valve and, on the suction stroke, takes in an equal amount through the suction check valve.
- *Double Disc Diaphragm* — Redundant diaphragms eliminate the need for contour plates.
- *Tubular Diaphragms or Bellows* — Uses a tube-shaped diaphragm.

*Primary Advantages* — discharge pressure: to 250 psig (mechanical design) and up to 4000 psi (hydraulic version); chemically inert wetted parts; adjustable stroke and RPM

*Limitations* — presence of check valves (balls, springs, o-rings); diaphragms require routine maintenance

The following are examples of diaphragm pumps:

## Gear Pumps for Metering Applications

In gear pumps, one of the gears is turned by a power source and drives the other gear(s). The spaces between the gear teeth carry the fluid from the inlet to the outlet ports. The gear mesh point prevents the fluid from returning to the inlet side. Typically, the drive is magnetically coupled to the gear mechanism (see Figure 48).

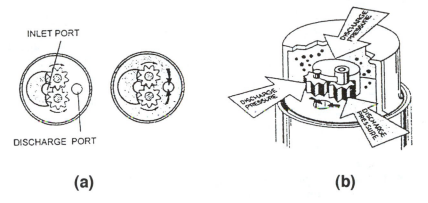

INLET PORT

DISCHARGE PORT

DISCHARGE PRESSURE

DISCHARGE PRESSURE

DISCHARGE PRESSURE

**(a)** **(b)**

**FIGURE 48** Gear pumps for metering applications.
(a) cavity style pump head
(b) pressure-loaded pump head.

*Primary Advantages* — pulseless flow; magnetically coupled drive eliminates seals; differential pressure: 100 psi; system pressure: 300 psi (typical)

*Limitations* — liquids with particles and slurries cause gear problems; self-priming only about 1 ft lift

The following are examples of gear pumps used for metering applications:

- *Cavity-Style Pump Heads* — They depend on the surrounding pressure in the magnet cup cavity to hold the gears tightly together. Since it is imperative that the gear tips mesh exactly, cavity style pump heads often use helical gears composed of extremely low friction material, such as Teflon, as a good alternative to stainless steels when galling is an issue, as well as for moderate pressure applications.
- *Pressure-Loaded Pump Heads* — They increase volumetric efficiency and use a suction shoe which works with the fluid pressure in the magnet cup to seal the gears tightly together. As the discharge pressure on the suction shoe increases, so does the efficiency of the pump. The pressure on the outlet side of the pump should be greater than the pressure on the inlet side in order to force the suction shoe and gears together.

## Peristaltic Tubing Pumps

In a peristaltic tubing pump, the pump head consists of two parts: the rotor, which contains three or more tubing rollers, and the housing. The tubing is placed in the "tubing bed" between the rotor and housing. The rollers on the rotor move across the tubing, squeezing it against the housing and pushing the fluid. The tubing behind the rollers recovers its shape, creates a vacuum, and draws fluid in behind it. A pocket of fluid is formed between the rollers. This volume is specific to the inside

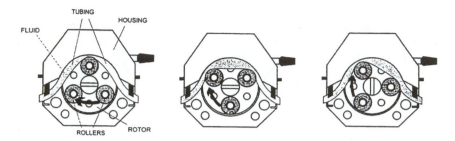

**FIGURE 49** Peristaltic tubing pump for metering applications.

diameter of the tubing and the geometry of the rotor. Flow rate is determined by multiplying speed by the size of the packet (see Figure 49).

> *Primary Advantages* — No mechanical wetted parts (other than tubing); no seals, packings, or check valves; easy to clean (dispose of tubing); multiple pump head configurations
>
> *Limitations* — Pressure is under about 40 psig; routine tubing replacement (depends on tubing material, fluid, drive RPM, and required accuracy); tubing life is rated by either failure (leak) or a decrease in accuracy of 50%; therefore, tubing life is over-rated for metering applications; suction lift is limited to tubing wall strength

### Rotating/Reciprocating Valveless Pumps

Rotating/reciprocating pumps are no-packing pumps, with extremely close clearances between piston and matched liner to eliminate packings. The capillary action of the fluid effectively acts as the primary piston seal. The absence of packings and the use of sapphire-hard ceramics virtually eliminate periodic maintenance associated with packed plunger type pumps. Rotating piston design uses a unique piston to accomplish both pumping and valving, eliminating valves used in traditional piston pumps (see Figure 50).

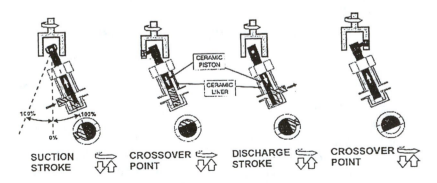

**FIGURE 50** Rotating/reciprocating valveless pumps for metering, dosing, and lab applications.

The piston rotates and reciprocates. As the piston is pulled back and the piston flat opens to the inlet port, suction is created, and fluid fills the pump chamber. The inlet port is then sealed and cross-over occurs. As the inlet port is sealed and the pump chamber is full, the outlet port opens up. Only one port is open at any time, and at no time are both ports interconnected. The piston is forced down and the piston flat opens to the outlet port. Discharge is created, and fluid is pumped out. The position bottoms for maximum bubble clearing. When the outlet port is sealed and the pump chamber is empty, the inlet port opens to start another suction stroke.

*Primary Advantages* — no valves to clog, hang up, or service; no packings or seals which require routine maintenance; one moving part; self-priming to 15 ft lift; dimensionally stable ceramic internal parts

*Limitations* — the process fluid is the seal and lubricant; flush gland required for air-sensitive fluids

## METERING PUMPS SELECTION

Metering pumps are used in a very diverse range of applications. Specifications for metering pump applications generally consider the following parameters:

*Volume* — Flow rate or volume per dispense.

*Pressure* — Operating pressure, system pressure, differential pressure (pressure across the inlet and discharge ports).

*Accuracy* — Accuracy, precision (repeatability).

*Temperature* — Process temperature, ambient temperature, non-operating temperature (component sterilization).

*Plumbing* — Connection size, material, fitting (Example: 1.5', 316 SS, sanitary). Suction side tubing size is critical to preventing cavitation, which may occur when the tubing size is too small to allow adequate flow of fluid to enter the pump. Cavitation can cause problems including hammering, flow variation, bubble formation and loss of prime, and eventually pump damage.

*Drive Power* — AC, DC, pneumatic, stepper, mechanical.

*Flow Control* — Fixed, variable, reversing.

*Control Source* — Manual, electronic (Example: 4-20 mA loop), mechanical.

*Special Requirements* — Hazardous (X-Proof), sanitary (3A Approved), sterile, outdoor.

The user also needs to specify and address the following: fluid characteristics, chemical compatibility (with pump wetted parts or fluid path), viscosity, corrosive application (pH = ?), specific gravity, suspended solids (percent and particle size), air sensitivity, crystallizing characteristics, shear sensitivity, size, and price.

## CONTROL AND INTEGRATION

*Manual Flow Control* — Utilizing drive mechanism adjustments or manually adjusting drive speed.

*Mechanical Flow Control* — Mechanically links an external drive source to the pump head (Example: Directly linking a speedometer cable to a pump head for agricultural spraying applications).

*Electronic Flow Control* — (See Figure 51).

    *Open and Closed Loop Control Systems* — Direct electronic feedback from a sensing device (pH sensor, flow meter, level switch) to an electronic input on the pump (Micro-switch, RS232, 4-20 mA, 0-1 VDC, 0-10 VDC) (see Figure 51).

    *Programmable Logic Controller (PLC), Computer (PC)* — Integrate electronic protocols used in closed loop systems with software programs designed for process control. PLC- and PC-based systems are capable of monitoring several input signals and controlling the associated process functions (pumps, valves, heaters, mixers) simultaneously.

*DC Servo and Stepper Motor Controls* — Often used for analytical and medical instrumentation, it is usually integrated into the design of the instrument's electronic circuitry, although a variety of servo and stepper controllers are commercially available.

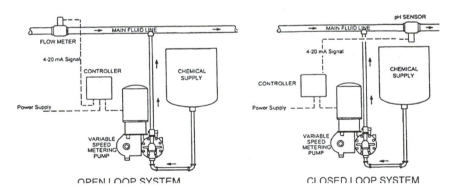

**FIGURE 51** Control systems for metering pumps.

## ACCESSORIES

*Pulsation Dampeners* — The pulsations generated by reciprocating metering pumps produce pressure that the system sees in the form of inertia, acceleration, shock, noise, and reduced service life. Placed on the discharge side of the pump, the pulse dampener (see Figure 52) acts as a shock absorber and makes the flow more uniform. On the suction side, the pulse dampener acts as an accumulator, reducing cavitation conditions. Most pulsation dampeners consist of a chamber containing a diaphragm or bellows that separates the process fluid from pressurized gas. With each pulse, the gas will compress and relieve a portion of the pressure generated. Devices as simple as corrugated in-line tubing can also be used in many applications to reduce pulsation.

*Strainers and Filters* — Remove solid particles from fluid stream that could damage internal pump components that usually have very tight clearances.

*Check Valves* — Incorporated into most metering pump designs, additional check valves are often added to increase pumping efficiency by preventing backflow. They also help maintain pressure on the suction side of pressure-loaded gear pump heads.

*Pressure Relief Valve* — Since metering pumps are positive displacement pumps, a pressure relief valve is often considered for the discharge piping to relieve extreme pressure conditions, thereby protecting the process instrumentation, piping, and in most cases, the pump itself.

*Flow Detection Device* — Most pumps rely on the presence of fluid for both cooling and lubrication. Flow detection devices are often installed to detect a no-flow or dry condition on the suction side to protect the pump from running dry. A provision must be made, usually in the form of a time delay, to allow the pump to prime, or not to be falsely shut down by small volumes of air.

*Pump Head Heating Devices* — Usually steam, hot water, or electric, these devices heat the pump head, allowing viscous fluids to flow more easily.

*Isolation Glands* — Used to isolate air-sensitive process fluids from seals and packings.

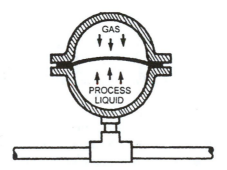

**FIGURE 52** Pulsation dampener.

Other auxiliaries may include back pressure valve, pressure gauges, etc.

## TYPICAL APPLICATIONS

Industrial:

*Water and Wastewater Treatment* — addition of sanitizing agents, pH control and conditioning

*Chemical Process and Instrumentation* — catalysts, additives, paints and pigments, etc.

*Agricultural Spraying, Chemigation, and Mosquito Control* — metering pesticides, herbicides, nutrients into fluid lines; grain storage preservatives

*Metal Stamping, Finishing, and Casting* — stamping lubricants, coatings, foundry catalysts

*Precision Cleaning* — parts washing, laboratory glassware washers, car washes

*Battery Manufacturing* — dispensing electrolytes, slurries, and stamping lubricants for button cell batteries

*Semiconductor Manufacturing* — dispensing of photo chemicals, developing and etching solutions, wafer cleaning, slurry metering

*Food* — addition of ingredients, colors, flavors, preservatives, and coatings for candy

*Tobacco Manufacturing* — controlling moisture content and flavor (menthol) addition

*Pharmaceutical Manufacturing* — laboratory, process chemistry, and packaging (filling and dispensing)

Analytical Instrumentation:

*Preparative Chromatography*
*On-Line Process Analyzers*
*Air and Water Sampling Analyzers* — emissions from factories (stack gas)
*Titration and Dissolution Apparatus*
*Spectrometers*

Medical:

*Diagnostic Equipment and Clinical Chemistry* — automated in-vitro diagnostics

*Dialysis Equipment* — precision re-circulation of dialysate

*Contact Lens Manufacturing* — soft lens filling fluid

*Pharmaceutical and Biomedical Product Mfg.* — manufacturing and packaging

*Disposable Kits* — adhesive dispensing for plastic medical components using UV curable adhesives and solvent welding

# 9  The Advantages of Rotary Pumps

The advantages of rotary pumps are the following:

*Efficiency* — 20 to 5% (depending on pressure) higher efficiency for most typically pumped liquids, with high viscosities.

*Viscous Fluids Handling* — Above approximately 300 to 1000 SSU (such as DTE light oil), a centrifugal pump simply cannot be used, as viscous drag reduces efficiency to nearly zero. PD pumps continue to pump at high efficiency, with no problems.

*Pressure Versatility* — As an example, a $1^1/_2 \times 1$ to 6 centrifugal pump can only produce approximately 20 psi system pressure at 1800 RPM. A similar PD pump goes to much higher pressures. This goes back to the inherent characteristic of PD pumps as "flow generators," practically unaffected by pressure, within a wide range.

*Self-Priming* — a typical ANSI-dimensioned centrifugal pump that is commonly found in many chemical plants cannot lift liquid. A standard PD pump, such as a gear pump, can easily lift liquid in the range of 1 to 20 ft. Centrifugal pumps can be made self-priming, by adding a priming flood chamber in the inlet, however, this adds expense and violates dimensional interchangeability.

*Inlet Piping* — Centrifugal pumps are extremely sensitive to inlet piping details. Improper piping may cause an increased NPSHR, cavitation, high vibrations, and possible damage to seals and bearings. PD pumps are less sensitive to inlet piping and can be a real solution for many difficult installations with space constraints, since piping modifications to the existing setups are very costly.

*Bi-directional* — By simply reversing the direction of motor rotation, many PD pumps will pump in reverse, which can be advantageous in many processes. Centrifugal pumps can pump in only one direction. In some installations, two centrifugal pumps are used: one for loading, and another for unloading, which doubles the piping runs, valves, and auxiliaries. A *single* rotary pump would do the same job. (Note: Relief valves should be installed in both directions in such cases.)

*Flow Maintainability* — PD pumps produce almost constant flow, regardless of fluids' properties and conditions (viscosity, pressure, and temperature). For centrifugal pumps, a change in fluid properties and external conditions would result in a definite change in performance.

*Metering Capability* — PD pumps can be used as convenient and simple metering devices. Centrifugal pumps have no such capabilities.

*Inventory Reduction* — Since PD pumps can pump a wide variety of fluids in an extreme range of viscosities, the same pump parts inventory is required for a wide range of applications throughout the plant. Centrifugal pumps require a greater multitude of sizes for different applications, which results in increased inventory of parts.

Nevertheless, centrifugal pumps also have numerous advantages and operate very well, especially at very low viscosities (such as water, etc.) when applied properly, as will be seen from the examples that follow.

# 10 Case History #1 — Double-Suction Centrifugal Pump Suction Problems

The job of a centrifugal pump is to generate enough pressure to overcome system hydraulic resistance. But it is the "responsibility" of the *liquid* to get to the pump. If the suction pressure is too low, the hydraulic losses can "eat away" enough pressure so that it could drop below the liquid vapor pressure. (A simplification of boiling mechanism with reduced pressure is shown in Figure 53.) Such formation of vapor bubbles (cavities) is called cavitation. If there is a lot of bubbles in the impeller inlet, the impeller may become "vapor locked," and the pump will stop delivering flow. If cavitation is less severe, a partial loss of flow will occur, accompanied by noise, pulsations, vibrations, and pump failures.[23]

As bubbles progress through the impeller, they enter the area of increased pressure. The bubbles then collapse with high intensity, as sort of micro-explosions. Some of these bubbles collapse within and into the surrounding liquid, causing micro-shocks (high frequency pulsations), and those that happen to be in close proximity to impeller walls (blades, shrouds) will transfer their "micro-hammer"-like impulses to the metal itself, eventually eroding it. These erosions can become very substantial and can eat away big chunks of blades, especially in the inlet region (Figure 54).[24] This is another reason why this suction phenomenon is called "cavitation" (i.e., blades being cavitating away). The best way to fight cavitation is to ensure that the suction pressure is adequate for the pump and for the particular liquid it is pumping.[25]

Often, the suction pressure is expressed in feet of head above the vapor pressure (also expressed in feet), with adjustment for losses (see Figure 55). Obviously, the available NPSH (NPSHA) must be greater than that required by the pump (NPSHR):

$$NPSHA > NPSHR, \text{ or } NPSHA = NPSHR + \text{ safety addition.} \quad (43)$$

The NPSHR is a characteristic of a pump and is determined by testing at the manufacturer's test facilities. During the test, the suction pressure is lowered until

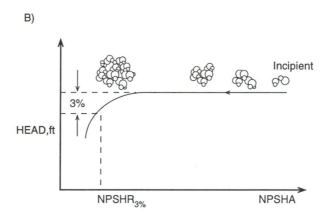

**FIGURE 53** Concept of incipient cavitation, bubble formation and growth, to full boiling.

the pump flow begins to drop. The HI has defined the NPSHR as when the pump head has decreased by 3%, from the stable condition. There is often controversy about what stable condition is, as we shall see later.

There are two ways to drop suction pressure during testing: by evacuating the suction tank or by throttling the suction valve in front of the pump. The throttling is simpler, but the valve causes flow disturbances, propagating toward and into the pump. These disturbances cause the pump head to drop sooner than it would in case of a more uniform (unobstructed by the valve) flow (i.e., the tested NPSHR value would be conservatively, or artificially, high). The vacuum method is more "pure," and produces more "scientific" results, unobscured by the presence of the valve. There are arguments, though, that the valving method is better because it represents the reality of actual field installations that always have some obstructions in the inlet and must be accounted for. So, an extra margin, obtained via a throttling test, may actually be desirable from the conservative standpoint.

The NPSH tests are conducted at various flow rates, and results are cross-plotted as head vs. NPSHA curves, as well as NPSHR vs. flow curves (see Figure 56).

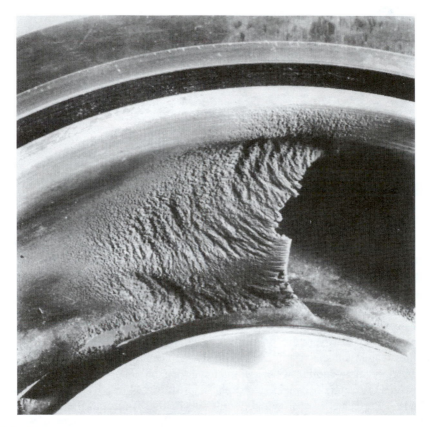

**FIGURE 54** Cavitation damage of the impeller blades in the inlet region.[24]

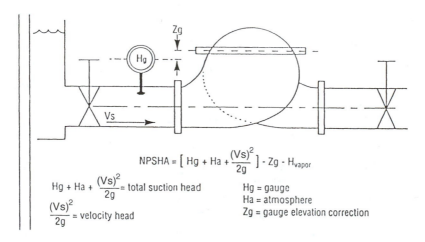

$$NPSHA = \left[ Hg + Ha + \frac{(Vs)^2}{2g} \right] - Zg - H_{vapor}$$

$Hg + Ha + \dfrac{(Vs)^2}{2g}$ = total suction head

$\dfrac{(Vs)^2}{2g}$ = velocity head

$Hg$ = gauge
$Ha$ = atmosphere
$Zg$ = gauge elevation correction

**FIGURE 55** Concept of NPSHA, with defining calculations.

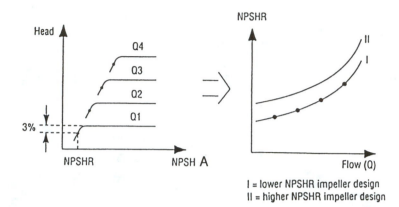

I = lower NPSHR impeller design
II = higher NPSHR impeller design

**FIGURE 56** Defining NPSHR$_{3\%}$, based on 3% head drop as NPSHA is reduced.

It is important to make sure that the user has sufficient NPSHA above the NPSHR at the complete *range* of flows, and not simply at the BEP or rated points (see Figure 57).

An example of a bad suction situation for a large double-suction cooling water pump installation is discussed in the following paragraphs (see Figure 58). These pumps experienced hydraulic noise over 90 dBA, and the life of the cast iron impellers was less than six months; obviously the customer was not very happy.

Typically, large cooling water pumps have their NPSHA in the range of 30 to 40 ft, which is essentially equal to atmospheric pressure (34 ft) plus the submergence of the impeller centerline below the liquid level (2 to 5 ft), minus the losses in the suction piping (5 to 6 ft). It is always important to make sure there is enough submergence, not only from the point of higher suction pressure to prevent cavitation, but also from the priming ability standpoint. Some cooling water pumps are installed with the "lift" condition, where the liquid level is below the impeller centerline (6 to 10 ft), and the checkvalve is required for priming. A checkvalve is sometimes prone to leaking, and it also adds to losses in suction line, thus decreasing NPSHA further. With flooded suction, starting the pump is much simpler, and makes an operator's job easier. In this case, the initial problem assessment and factory and site testing took almost six months, with no findings pointing to any NPSH problem. In other words, NPSHA appeared to be adequate, as compared to NPSHR, and should not have caused a cavitation problem (see Figure 57).

The R-ratio (NPSHA/NPSHR) is defined as NPSH margin. The reason to have a good NPSH margin is to ensure as much "bubble-free" suction flow as possible. Consider the illustration in Figure 53. As suction pressure lowered (NPSHA reduced), initial bubbles begin to form at significantly higher NPSH than the NPSHR$_{3\%}$.

You can verify this occurrence by conducting a simple experiment at home, by observing the boiling of water in a transparent pan. At first, tiny bubbles appear, and eventually a complete "percolating" process follows. The same happens in a pump, except that boiling is induced not by adding heat at constant pressure, but by lowering pressure at constant temperature. Cavitation (formation of bubbles), therefore, develops over a range of suction pressures, and is not instantaneous. Certainly, it would

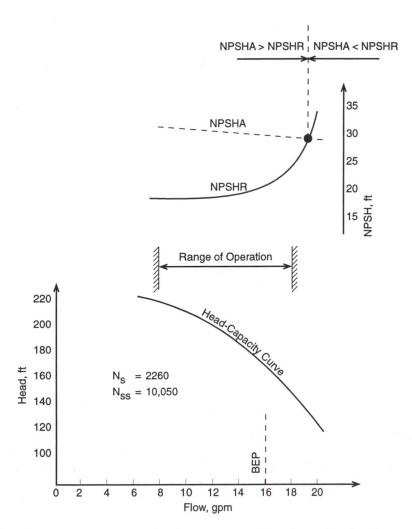

**FIGURE 57** Within the pump operating region, the NPSHA margin should not have caused problems in this example. So, why are there problems?

be 'nice' to have NPSHA > $NPSH_i$ (i = incipient, with no bubbles at all), but this is impractical because the R-value would need to be in the range of R = 10 to 20, in order to achieve nearly bubble-free flow. Besides, each foot of extra NPSHA would require increased construction costs, since the cooling tower would need to be raised. These economic considerations have led historically to typical R-values of 1.2 to 1.6 for cooling water pumps.

It is known from the previous field and lab studies on cavitation that the maximum damage to the impeller occurs not at the $NPSHR_{3\%}$, as might be intuitively expected, but at somewhere between that value and $NPSH_i$.[26,27] In the region of incipient cavitation, bubbles are too small and too few, and most of them collapse into the liquid which surrounds them. The hydro-acoustic noise in that region can

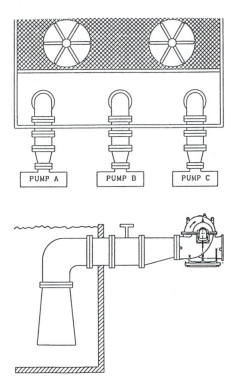

**FIGURE 58** Double-suction cooling tower water pumps installed at major chemical plants.[28]

be detected only with the use of specialized equipment. As pressure is reduced further, more bubbles form, and noise, pulsations, and vibrations quickly begin to rise. At some point, there are so many bubbles in the stream, and they are so large, that some of them begin to cushion the collapsing impulses of others. The noise is reduced, while the pump still continues to pump (see Figure 59). Further reduction in pressure produces a great amount of bubbles to the extent that the suction passages become vapor-locked, and choke the flow — the pump stops producing head and flow, and it essentially is running dry. Lowering the pressure beyond this region extends the cloud of bubbles from the impeller eye upstream into the suction pipe, causing a very severe flow regime — intermittent chunks of liquid, followed by vapor, then again by liquid slugs, and so on — a very unstable and violent process, with high vibrations, noise, and damage to the equipment. The nature of this type of noise is different: a more "metallic" sound, rather than the "pebble-like" sound of "classic" cavitation. ***You better get the pump out of this region fast!***

As was explained earlier, the best way to avoid cavitation is to go as far as possible into the region "R" (to the "right" of the maximum damage point on a curve, see Figure 60), thereby into higher suction pressure range. For many small process pumps, such as ANSI, API,[29] etc., this is not too difficult to accomplish. For example, a $1.5 \times 1 - 8$ ANSI pump running at 1800 RPM may require, say, 5 ft NPSHR, and a 10-fold margin ($Ri = NPSH_i/NPSHR = 10$), would mean 50 ft

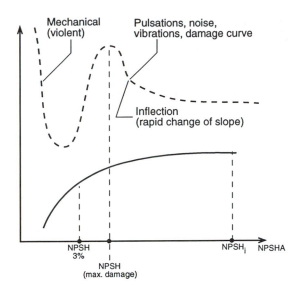

FIGURE 59 "Damage curve," as a function of NPSHA.

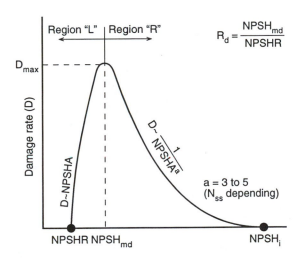

FIGURE 60 "D-curve," damage vs. NPSHA

(or approximately 20 psia ≈ 5 psig) suction pressure. This is easy to achieve for small pressurized tanks. However, for large pumps, such as cooling tower water pumps, there is no economic way to increase suction pressure, and some other methods to fight cavitation are needed. Since it is not possible to move away from the maximum damage zone into the "R" region, perhaps a shift into the "L" region could help. There is little room to maneuver, however, in order to fine-tune the pump in that region, since the "L" region is rather small and its boundaries are not known a priori: going too far will get into a fully cavitating region with considerable loss of flow, and not going far enough does not solve the damage problem significantly.

First, we need to pinpoint the "L" region more accurately, and the shape of the noise curve must be determined. The simplest way to accomplish this is with a hand-held sound meter, although measuring pulsations and/or vibrations would work as well. Pulsation measurement is a somewhat specialized process, requiring transducers, power supply, and processing equipment. Vibrations, on the other hand, are simpler, but may not produce enough magnitude variation between the data points to pinpoint with sufficient accuracy the slope of the changes of what is going on inside the pump. Sound measurements are easy.

As can be seen from Figure 59, the shape of the noise curve has an inflection where the slope changes to a more rapidly rising curve. This is known as an onset of recirculation, as famous papers by Frazer[30] and Karassik[31] explain. The onset of recirculation modifies the NPSHR characteristic. Suction recirculation is *not* desirable. A mechanism of suction recirculation is shown in Figure 61.

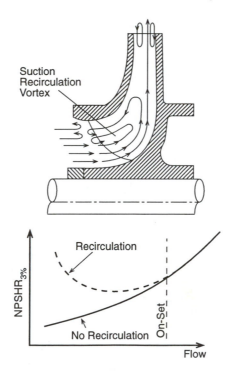

**FIGURE 61** Impeller eye recirculation increases NPSHR$_{3\%}$. More NPSHA is required to compensate.

A significant effect on the recirculation can be made by changing the impeller eye size — the larger the eye, the more prone is the impeller to recirculation (see Figure 62). However, if the eye is too small, the fluid velocity entering the impeller eye is increased, and (back to Bernoulli) the higher velocity causes static pressure to decrease, thus raising NPSHR.

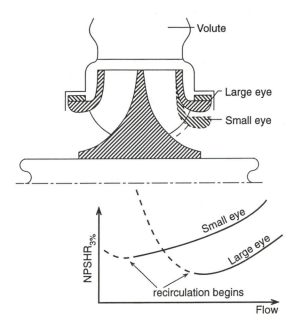

**FIGURE 62** Large eye allows less NPSHR (and, therefore, NPSHA), but pump operation is at limited flow range. Small eye requires more NPSHR (NPSHA) at the BEP margin, but allows operation at lower flow due to preventing (delaying) recirculation.

Let us now introduce the concept of suction specific speed (Nss):

$$\text{Nss} = \frac{\text{RPM}\sqrt{Q}}{\text{NPSHR}^{0.75}}. \tag{44}$$

Equation 44 is similar to Equation 34, except that the head is substituted by the NPSHR *at the BEP.* Just as Ns characterizes the "contour" of the impeller mostly at the discharge region, the Nss characterizes its suction (eye) shape (see Figure 63).

Typically, the range of Nss varies from 7000 to 15,000, with special designs (inducers) of even higher values. Through experience, mainly from the pump users' community (Amoco Oil, DuPont, and others), it has been found that over approximately Nss = 9500, the rate of pump failures attributed to suction increases significantly (see Figure 64). The HI has suggested even a more conservative design value, of Nss = 8500, which limits the eye size.

It is true that if a given pump was never to operate outside its BEP point, a much higher value of Nss could be used (and is even desirable), resulting in lower NPSHR (see Figure 62), and therefore allowing lower NPSHA, thereby lowering construction cost. There are some rare cases when such designs are made, such as in an existing installation with inadequate NPSHA, where the low NPSHR impeller is the only solution, provided that the range of flows is very narrow and stays around the BEP. In most cases, however, centrifugal pumps operate in a wide range of flows, below and past the BEP point, reflecting the variable plant need for cooling water. When

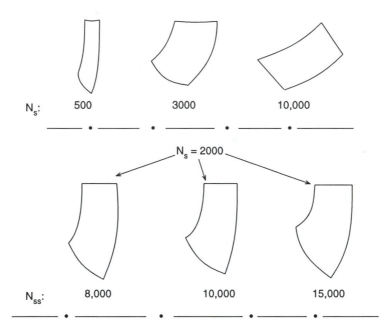

**FIGURE 63** Specific speed (Ns) characterizes the overall impeller profile shape. Suction specific speed (Nss) characterizes the eye size, for given specific speed (Ns).

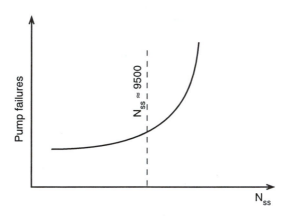

**FIGURE 64** Industrial pump users' experience: number of failures tend to increase very rapidly above Nss ≈ 9500.

operating at low flow, a recirculation becomes a reality, and the lower Nss designs could become "life savers," as they are more forgiving in that regard. This is why it is customary to limit the designs to Nss = 9000 to 10000. (Note: There is some confusion in deciding which flow to use to calculate the value of Nss for double suction pumps. In some papers, this flow is taken as half total pump flow, i.e., flow per eye.) The HI, however, maintains a definition of Nss based on total pump flow.

For this reason, when comparing Nss for different designs, make sure you know which flow was used for the Nss definition).

## QUIZ #5 — ARE Ns AND Nss DEPENDENT ON SPEED?

Show that Ns defines the impeller contour only by hypothetically speeding up your pump. Will Ns change? Why?

### SOLUTION TO QUIZ #5

$$N_s = \frac{N\sqrt{Q}}{H^{.75}}, \quad N^*_{(new)} = KN \ (K = \text{speed increase multiple}).$$

By Affinity,

$$Q^*/Q = K, \qquad H^*/H \sim K^2,$$

so:

$$N^*_s = \frac{(KN)\sqrt{KQ}}{\left(K^2 H\right)^{.75}} = \frac{K^{3/2}NQ^{1/2}}{K^{3/2}H^{3/4}} = \frac{N\sqrt{Q}}{H^{.75}} = \text{const.}, \quad N^*_s = N_s = \text{const.}$$

To return to our case, since we could not find anything wrong mechanically, and the NPSHA also appeared to be adequate, suction recirculation became suspect, and two design modifications were made. First, a smaller eye impeller was designed and tested in the field. As can be seen from Figure 65, the smaller eye indeed suppressed the onset of recirculation toward lower flow, and the sound curve has shifted to the left (toward lower flow), thus making the sound level lower for most of the operating region.

As would be (unfortunately) expected, the pump now could not, however, operate as far out to higher flows beyond the BEP because of NPSHR "stonewalling" at high flow, quickly exceeding the NPSHA in that region. Fortunately, the plant concern was in the lower region, where this pump needed to operate, and coverage for higher flows was not important. The solution, therefore, was considered an improvement.

Upon closer examination of Figure 65, the sound level appears to be reduced even more than simply by the curve shift (toward the low flow) — as if by some additional effect. The reason for this additional improvement in noise reduction is that the eye size was not the only design change made for suction performance improvements. The actual shape of the impeller blade inlet tips is also very critical (Figure 66). Analogous with an airplane wing,[32] a blunt and badly contoured blade leading edge causes vertices, flow separation, turbulence, and loss of performance. The shape of the impeller inlets is important. A proper contouring of the inlets can improve suction characteristics and impeller life very significantly. Also, a proper

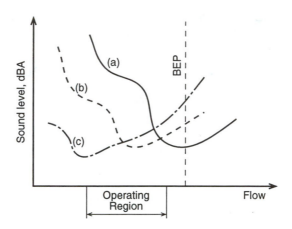

**FIGURE 65**  (a)  Original large eye impeller
(b)  Expected curve shift, with smaller eye
(c)  Actual improvement, including combined effects of smaller eye, better blade inlet profiling, and casing suction splitter modifications (contouring)

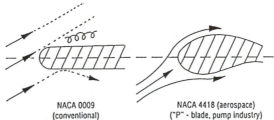

NACA 0009                  NACA 4418 (aerospace)
(conventional)              ("P" - blade, pump industry)
Blade inlet optimization, using airfoil shaping to desensitize inlets to flow incidence,
i.e. better cavitation characteristics.

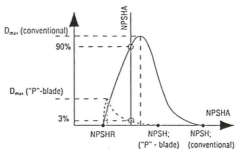

Reduction in cavitation damage due to blade profiling ("P" - blade).

**FIGURE 66**  Impeller blade inlet profiling improves suction performance.

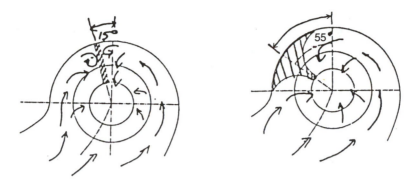

**FIGURE 67** Comparison between the old (left) and new (right) splitter design for better flow feed into the impeller eye.

design of the suction splitter (see Figure 67) of the casing can further improve flow uniformity and suction performance.

Therefore, the new design combined all three improvement features: smaller eye size, a so-called "P-blade" contour, and a casing suction splitter modification (better profiling). Low flow separation was much deterred and improved, which explains additional noise reduction benefits.

These, and similar troubleshooting techniques, can be applied to "stubborn" cooling water pumps. In each such case, you need to examine the installation carefully, gather data, and evaluate the design in terms of inlet conditions. A combination of analysis, modifications, modeling, and testing may be required. Usually, the choice of the recommended actions is wide, and it is important to pinpoint the most effective and economical ways to solve each particular pumping problem.

# 11 Case History #2 — Lube Oil Gear Pump: Noise and Wear Reduction

Next, we will examine the troubleshooting methods for the case of an external gear pump operating at 415 RPM (3.55:1 gearbox reduction from 1470 RPM motor speed), delivering 219 gpm of heavy oil (2500 SSU) in the blending plant. The pump operation was reported to be noisy, and the customer contacted the pump supplier, after a local environmental inspector reported his findings on noise levels as exceeding the allowable limit and in need of correction.

First, the alignment of the endplate and faceplate to the casing was checked, to ensure that the dowel pins' location was correct — otherwise, the bushings "misposition" would offset the shafts (with gears on them) onto the case, causing rubbing and, therefore, noise and wear. Gear pumps are often made from cast iron, which is known to have good self-lubricating properties due to high carbon content. This ensures non-galling operation, even if gears, deflected by the pressure differential, contact the casing. While not always causing catastrophic failures, this could result in noise and premature wear. The inspection of the pump components had shown that the parts were machined correctly and the dowels were positioned properly, so that no mechanical contact between gears and the case should be occurring (see Figure 68).

The design of a gear pump is typically such that the bearings' clearance is less then the clearance between the gears and case. This is very important for stainless steel pumps (especially 316ss material), particularly when pumping thin liquids, but even in the case of iron construction, it is still a good design practice to follow the rule of avoiding mechanical contact (which appeared to be true for this problem pump case).

## QUIZ #6 — DOUBLE GEAR PUMP LIFE IN TEN MINUTES?

Can a worn-out gear pump life be doubled with no new parts and modifications?

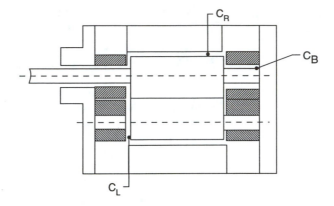

$2C_R$ = diametral clearance between the gear and case
$2C_L$ = total lateral clearance, gear/endplates
$2C_B$ = diametral clearance between the bushings and shaft journals

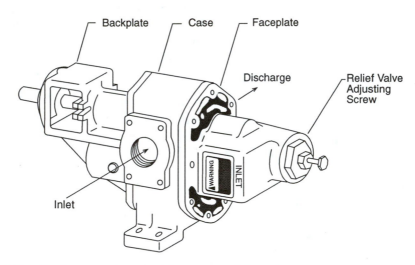

**FIGURE 68** Gear pump for Case #2. (Courtesy of Roper Pump.)

## SOLUTION TO QUIZ #6

If a gear pump casing is turned around, for example, as discharge and suction ports are swapped, the worn-out part of the casing would now face the discharge end, and the non-worn section would face toward the suction end. The wear of the casing is caused by the differential pressure pushing (deflecting) the shafts, with gears, onto the casing wall. The force moves the gears away from the discharge portion of the casing, and toward the suction, where wear takes place. When the casing is turned around, the non-worn part of the casing will actually restore the radial clearance between the gears and casing, thereby reducing the slip, and restoring the pump flow. Certainly, this solution is not a permanent fix, for the casing and gears will continue to wear, but it will prolong the operating life while a new pump is procured (see Figure 69). The main problem, however, is that the pump is probably misapplied to begin with: the differential pressure must be reduced, or a higher pressure rated pump design should be installed.

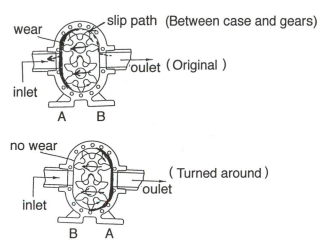

**FIGURE 69** Increasing life of the worn-out pump.

Continuing on with our problem. Because no obvious *mechanical* problems were found, the *hydraulics* was suspected next, and cavitation was considered, as explained below.

For rotary gear pumps, the measure of suction performance is minimum allowable suction pressure. This is different from the centrifugal pumps, where NPSHR (and not suction pressure itself) is used as a criterion. Similarly, with centrifugals, a certain percentage drop in suction performance is used to establish $p_{s_{min}}$, except that instead of the drop in head (as in centrifugals), a drop in flow is used. The HI defines a 5% decrease in flow as a threshold at which minimum required suction pressure is defined (see Figure 70).

For multicomponent liquids, which are indeed often pumped by the gear pumps (i.e., complex oils), the shape of the curve may not have a well-pronounced drop, but a more gradual, sloping decrease ("droop"). The reason for this is a "step-like" vaporization of the components (e.g., aromatics) as pressure is reduced. Data about minimum required suction pressure is available from the manufacturers and should be carefully considered when selecting a pump. This data is based not only on cavitation aspects in a "classic" sense (i.e., vaporization, or bubble formation), but also should include considerations of the effect of a different nature — that is the ability of the incoming liquid to *fill the cavities* between the gear teeth during the time these cavities are exposed to the incoming liquid during the revolution of the shaft, as was explained in the earlier section on gear pumps.

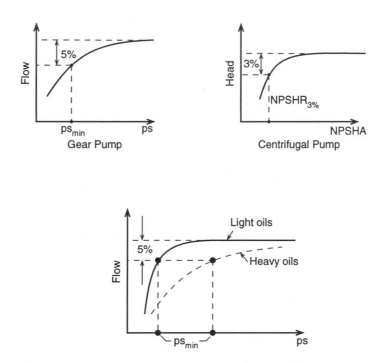

**FIGURE 70** Illustration of minimum allowable suction pressure for gear pumps, as compared with centrifugal pumps (HI definitions).

The $p_{s_{min}}$ criteria, known to a manufacturer, should be based on actual tests and not strictly on calculations, because the process of cavity filling is very complex, and depends on many factors, such as speed, viscosity, number of teeth, and casing porting arrangements.

For the pump in this example, the cavities between the gear teeth were not filled completely by the incoming liquid. The resulting vortices in voids generated a distinct noise, similar in nature to flow separation ("pebbles") in centrifugal pumps during recirculation.

Sometimes, the air injection into pump suction[33,34] may reduce the noise significantly. This is typically the case when cavitation has started but has not yet developed enough to the point of flow reduction below a significant level. As we recall from the discussion on centrifugal pumps' cavitation, the noise diminishes if the amount of vapor bubbles is significant enough to cushion the collapsing impulses of each other. The injected air essentially acts similar to vapor bubbles, resulting in a similar cushioning effect. This technique was tried with this pump, and (to everyone's initial satisfaction, the noise disappeared completely.

This, however, was considered only as a temporary solution: first, the air supply at the pump inlet was not readily available, and the presence of the compressor just for this purpose could not be justified. The customer was also concerned that the injected air may cause unknown and undesirable effects downstream, where it would tend to come out of the solution in large volume at the lower pressure regions.

Therefore, a method similar to the one discussed in the previous example with a centrifugal pump was tried. To drop suction pressure sufficiently, to enter region "L," the suction valve was throttled. The noise was reduced substantially as shown in Figure 71. When the valve is throttled even further, the noise would rise again, and much more violently now with a distinct metallic "clanking," as gears begin to run essentially dry — this is similar to the example with the centrifugal pump in the previous case. In other words, the valve position would have to be fine-tuned very accurately during the operation, in order to stay away from the high noise level, but not to the extent of choking off the suction completely. If the level in the supply tank would change, the inlet pressure in front of the pump would force the pump to operate either back at the high noise point of cavitation, or swing the other way, dangerously close to vaporlocking condition. Manual readjustment of the valve would thus be required each time the process variations resulting in the tank level changed, which would be often and unpredictably. To automate valve position and adjust tank level would be impractical and expensive.

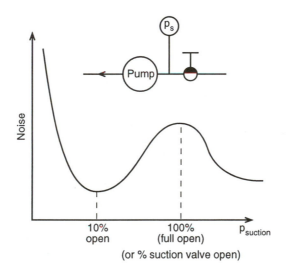

**FIGURE 71** Inlet valve position alters pump suction pressure, changing noise level.

As is often the case, the problem was certainly a combination of a pump and a system interaction — a different pump in the same system, or the same pump at a different system, operated satisfactory with no problems. In fact, the puzzling thing was the fact that the same design operated without problem in a multitude of other installations around the world.

In general, when looking at ways to improve a pump design, by making it less sensitive to suction conditions, a filling ability of the tooth cavity, as was explained in the earlier section on centrifugal pumps, is one of the most obvious areas of potential improvements. Many large gear pumps have small diametral pitch, i.e., few teeth for the pitch diameter. The number of teeth could be as little as six. If the number of teeth is increased, the cavity between the teeth gets smaller (just as a tooth itself gets smaller), making it easier to fill the cavity during the limited time of its exposure to suction port (see Figure 72). However, this has a significant impact on the design: if the same casing is to be used (by maintaining the same distance between gear centers, i.e., same pitch diameter), then the new pitch requires new hobs (cutting tools used to machine the gears), thereby adding cost. Besides, smaller pitch would produce smaller teeth, which would reduce pump capacity per each revolution (refer to the example in Quiz #4). If the design change is impractical and costly, a simpler way is to reduce the pump speed. The resident time of the cavity exposed to the suction port would increase, although the net flow would decrease. Sometimes the decrease in flow is acceptable (for example, in non-critical oil transfer applications), and this solution could reduce the noise substantially. This is what was done in this case; a gear reducer was retrofitted with a different set of transmission gears, and the pump speed was reduced from 414 RPM (3.55:1) to 284 RPM (5.18:1). The noise reduction was indeed substantial, and the solution was accepted by the plant.

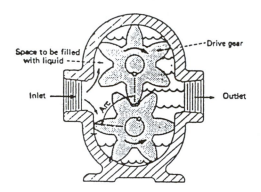

**FIGURE 72** Liquid must completely fill the space between the gear teeth to avoid cavitation.

Certainly, there are many situations where the reduction in flow would not be acceptable due to the process flow requirements, but for simple transfer services there is usually room to work. For cases where flow must be maintained (or even increased), a larger pump, running more slowly, would be a better initial choice. There are some other methods which can be considered, each of which must be analyzed on its own merits for applicability in each particular case, where seemingly insignificant nuances of the installation may turn out to be the Achilles heel that makes or breaks the success of the troubleshooting job.

Before jumping to quick solutions, it is important first to understand *all* of the variables and relevant factors, which can even *possibly* affect your pump and a system. Next, you must identify and decide which variables are critical, and plan the potential solutions to solve a problem, step by step. Often, if analysis and planning are done upfront, considerable time and effort could be saved overall in the long run.

# 12 Case History #3 — Progressing Cavity Pump Failures

As mentioned earlier, progressing cavity pumps (PC pumps) are known to operate very well in many particularly tough environments where abrasion and wear make other types of pumps impractical. In the case of this example, a medium-size PC pump was installed at the wastewater plant in the last stage of the treatment process. At that stage, all bacterial and other contamination impurities are already removed by the upstream processes (chemical additives, settling tanks, etc.). Now, the pump-age had to be transferred, as a final sludge of very high viscosity (500,000 cp), into large containers for trucks, to be subsequently hauled as fertilizer to the farm fields.

The problem reported was that the pump would stop pumping in less then one week. Rotors would wear badly, and increased leakage path would result, losing all of the pumped flow in slip. It was first suspected that the initial fit between the rotor and a stator was too tight, which, in the presence of a very dry viscous pumpage, would cause high friction and wear.

The relationship between the rotor/stator clearance and slip is not linear (see Figure 73). A good thing about the fit being tight is that there is practically no slip, especially if the number of stages is large at the same time. PC pumps are typically designed for approximately 75 psi per stage. A 300 psi application is typical for PC pumps, although there are extremes in both directions, with pressures as high as 2000 psi. For 300 psi, a 4–5 stage design would keep slip to a minimum. A bad thing, however, about a tight fit is high mechanical friction, resulting in high running, as well as starting, torque.

If fit is reduced, but still tight, the pump flow would remain relatively constant, and would not change much until the value of –0.015" diametral interference is reached ('–' indicates interferences, and '+' means clearance). Eventually, slip begins to take place, and increases almost linearly with change in interference, until about line-to-line fit is reached. After that, slip increases exponentially and very rapidly, and at +0.015"/+0.020" most of the flow could be recirculating back to suction, (i.e., fully slipped). The volumetric efficiency follows the trend as explained previously. (Note: the values are approximate and depend on other factors, some of which are explained subsequently).

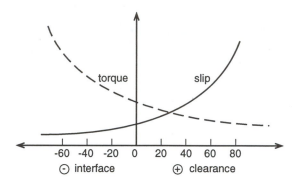

**FIGURE 73**  Slip and torque, as functions of rotor/stator fit.

The starting torque follows the reverse relationship as compared to slip. At first, with tight interference, the starting torque is extremely high — many field problems can be traced to this, and many motors have suffered when started in these (essentially locked rotor) conditions.

The trick is to know the correct interference or clearance, to satisfy both through-flow requirements with minimum slip, and to ensure that the pump starting ability is not impacted. The temperature is another factor to consider, since the elastomer (rubber) expands, changing the rotor/stator fit. It is probably better to err on the safe side, and end up with somewhat more slip, but longer life — the life of the rotor and stator decrease exponentially with fit.

Working with the customer, we did not find anything obviously wrong with the pump or the system; no blatant mistakes were made in the application. The factory test data and the design parameters did not point out at any obvious reasons for concern either. The pumpage was not abrasive, and all hard particulate inclusions had been removed way upstream before the pump. The pump was running at about 450 RPM.

After a full day of work, rather disappointed with lack of porgies, I stood some distance away from the pump, thinking about other possible causes of the problem. Absentmindedly, I picked up a handful of lime, which had seeped out of the bag laying nearby. Mauling it around in my hand, and thinking about the pump, my hand felt moisture apparently emanating from inside the lime. I could also sense sharp particles, like little stones, in my hand.

Lime is added to the pumpage in small quantities to "soften" the pumpage or to reduce the high viscosity of the sludge. Added at the pump suction, the lime produces a mild chemical reaction resulting in some water, which is helpful to lubricate the stator/rotor interference, to reduce friction (with the obvious intent to increase the pump life), as well as save energy cost via better efficiency.

We picked more lime, placed it in a container, and washed it out. The lime "melted away" and washed, but a few stones remained at the bottom of the container. The label on the bag indicated the lime contained only a very small percentage of impurities. Although the process sludge is not abrasive, the addition of the hard particulates with lime changed what was entering the pump suction.

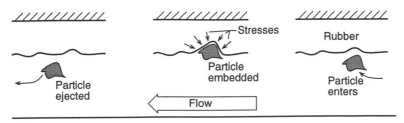

**FIGURE 74** Resistance to wear, due to rubber elastic energy to receive/eject particles.

Generally, PC pumps work well on abrasive applications, provided that the rotational speed is selected properly. When abrasives pass throughout the PC pump, it is the unique characteristics of the elastomers, such as rubber, that save the day: instead of fighting it, the rubber actually catches the particles, deflecting into itself, like a trampoline. This deflection results in elastic energy, which is then released to catapult (i.e., to eject) the particles out. Very little damage to the rubber results, but it can be spotted upon the examination as many small, often microscopic, cuts (see Figure 74).

When the clearance between the rotor and a stator is sufficient, there is enough space for the particles to be released to; otherwise, some of them get "jammed" just when the catapulting action is about to release them. If this natural swing-like mechanism of embedding-and-release is interrupted, or gets out of synchronism (similar to a swing abruptly stopped), the elastic energy is lost, and the particle remains embedded in the rubber. Depending on rotor/stator fit and particle size and shape, enough particles may become embedded in the elastomer, creating a very abrasive surface (see Figure 75). The wear that it causes is then obvious. Even if this lime was purchased at a premium with lower residual limestone, it would not take long to create the abrasive surface; perhaps one week failure rate could be stretched to two, but no more. The only way to handle these particulates is to remove them. Working together, we devised a simple cyclone separator, sort of a trap, to keep the particles from jumping over the wall of the "stopper" device we designed for the suction line (see Figure 76). Periodically (we calculated once a day), a hatch in the bottom of the trap would have to be opened and cleaned of particles — either manually or automatically.

The solution may be good for this relatively small pump size, but would it work for larger sizes? The inlet trap may or may not work for the larger sizes because the settling velocity of the injected lime could be different (if too high, it would carry the particles over the wall of the barrier), and again, some experimentation could be required. Often a simpler and possibly better solution, from the beginning, would be to apply a larger pump size, running at slower speed, perhaps 150 to 200 RPM. The slower speed would be more forgiving to abrasives. The somewhat higher initial cost could have been paid up in repairs, parts, and process downtime. The fit between the rotor and stator should also be loosened.

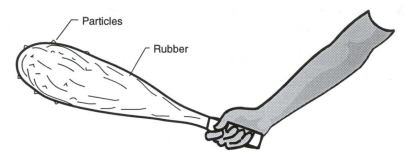

**FIGURE 75** A very abrasive surface.

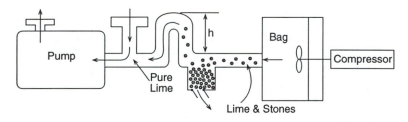

**FIGURE 76** Lime stone "stopper" device. Height "h" depends on particles' size, compressor air supply, and percent of allowable stone to the pump.

Overall, a combination and impact of several variables would need to be investigated and sorted out, perhaps with some testing, to devise a reliable, trouble-free, economical, and lasting pump operation.

Those who tried know that a solution to pump problems is often not obvious, and, just like the cases we discussed in this book, must be approached carefully and with consideration of all — even seemingly unimportant — factors.

# 13 Troubleshooting "Pointers" as Given by Interviewed Pump Users and Plant Personnel

- Rule Number One: "Always do the easy stuff first!"
- Always check the rotation of the pump.
- Make sure relief valve setting is correct.
- Check speed for belt-driven, or vari-drive, cases. Do not take them for granted.
- Check the integrity of the coupling.
- Alignment: How and who? Is piping forced to flanges?
- Trace the system and make sure the valves are open and that you are pumping from the correct vessel.
- For heavy viscosities, make sure the pump is warmed up prior to starting.
- Check for high temperature on the bearing housing to detect excessive thrust conditions. Bearings could tell you a good story.
- Inspect oil level in the bearing housing.
- Check for signs of discoloration.
- Inspect seals for leakage.
- Discuss changes in operating condition: New product introduction or adjustment? Compare stories.
- Discuss changes in line-up or changes in operating performance.
- Only after the above questions are resolved, consider the removal of the pump from service.

# 14 Application Criteria and Specification Parameters

The variety of pump types and sizes may be, and usually are, very confusing, and it is difficult even for the experienced pump users to make a proper/optimal selection, for his or her application. In many cases, a given plant simply follows the past practice in specifying, purchasing, and repairing of the rotating equipment. The requirements of various departments within the plant could also be conflicting at times, adding to the complexity of the decisions on pump application strategy. For example, low MTBF would require frequent repairs and a high maintenance budget, but the initial pump cost could be low and, in that sense, attractive to the purchasing group. A *total* cost would need to be evaluated, including cost of lost production during outages, spare parts, etc.

Another example could be an ease of flow control: It is easy to change the operating point of a centrifugal pump simply by throttling a discharge valve, and this could be attractive from the operator's viewpoint, especially when fast change in flow is critical to a process. However, flow control by valve throttling is inefficient and also causes high radial thrust (seals' and bearings' life reduced) at low flows or excessive NPSHR at high flows (cavitation damage and low impeller life). For a small, low horsepower pump, this may not be a critical issue, and operators' ease of control and flexibility would be a more important factor to consider. For high energy pumps, such as cooling water units, boiler feed and similarly, the savings in operating at or near the BEP vs. running the pumps off-peak, could be substantial. The investment into variable frequency drives might be a good option in such cases.

In reality, for the existing applications a change in a pump type, or even size, is usually difficult to implement for the above reasons, unless the problem is very pressing and serious. Technically, the change is usually not very difficult, but the implementation logistics prevent such changes. A centrifugal pump may have a very bad piping configuration at the suction side, causing cavitation, noise, and failures, but the realities of the existing space constraints may be such that piping changes are prohibitive, and the best approach well may be to just continue the existing policy and try to minimize the downside as much as possible (i.e., have spare parts (seals) on hand, conduct predictive maintenance to better pinpoint parts replacements before catastrophic failures occur, etc.).

For the new applications, the options are much more open. If a new facility is only at the design stage, this is the best time to investigate pumping options. The limitation of having to fit a different pump type or size into the existing piping is no longer a constraint in these cases, and does not require an additional investment (piping modifications to the existing installations can be very expensive, especially for larger pumps).

Referring to the HI pump chart (see Figure 1), the initial step is deciding on a pump type: kinetic, or a positive displacement. Although there are exceptions, as a general rule, **kinetic** pumps are used for *higher flows and lower pressures* (heads), while **positive displacement** types are used for *lower flows and higher pressures*.

## APPLICATION POINT #1: PUMP TYPE — SPECIFIC SPEED CRITERION (Ns)

$$Ns = \frac{RPM\sqrt{Q}}{H^{0.75}}$$

If Ns > 500, you are most likely in the area of centrifugal pumps.
If Ns < 500, a positive displacement pump is a likely candidate.

Keep in mind that a specific speed characterizes an individual impeller. For multi-stage pumps, a specific speed must be based on the head per stage, i.e., total head divided by the number of stages. For the initial selection, however, a particular pump type is not yet determined so the total head must be used to point out a pump type, in order to start off the selection process. As was mentioned earlier, there are exceptions to the rule. Some kinetic pump types can generate very high head (such as regenerative turbine pumps), and the specific speed criterion, as suggested above would not hold true for them. For the majority of cases, however, this criterion should give the user a reasonably reliable and quick method to determine the most likely solution to his or her application selection. If the user is aware of a reliable supplier of such special pumps, an additional inquiry may be worthwhile as it might turn out to be the best solution after all, for a given application.

## APPLICATION POINT #2 — WITHIN THE PUMP TYPE

Assuming the pump type is known, we need to begin to narrow down the pump type more specifically. If a centrifugal pump is selected, the choice is the overhung impeller, or between-bearings design. There is also a turbine type design, but the other selections are more straightforward and are traditionally reflected in the applications: space limitations of power plants and the need to take suction from the pit with relatively high flow of water, brine, or similar pumpage, agriculture, cooling water, etc.

Familiarity with the existing pump specifications is helpful, and a copy of these specs should be at the plant engineering department. Examples of major specifications are ANSI, typically used for applications at the chemical plants, and API, generally used for the refineries. Over the years, both ANSI and API specifications have grown to rather comprehensive documents, which now cover single-stage overhung centrifugal pumps (and more for API) and include dimensional requirements, sealing arrangements, materials of construction, etc. You can learn a lot about pumps, just from studying these specs and understanding the reasons for their particular requirements, and your training department may want to coordinate a short seminar for the plant operators, maintenance, engineering, and purchasing personnel. Such training is done either by internal experts or external consultants, and the time spent on such training is well worth the effort.

If you are going to pump water, for a very general application, a centrifugal pump is probably your best choice. It does not need to be an ANSI pump; you should base your decision on cost and reputation of the pump supplier. You would not go wrong with the ANSI pump, particularly if you envision changes in the future: dimensional interchangability of the ANSI pumps is a plus, making it easy to fit an ANSI pump made by any manufacturer into the same piping. The ANSI pump would cost a little more, but has the benefits mentioned before.

## APPLICATION POINT #3 — FLOW/PRESSURE

If a positive displacement pump type is considered, the reciprocating pump is the best choice for very high pressures, approaching 10,000 psi or more. For lower pressures, a rotary pump design could be selected. A "Rotary Pumps Coverage" chart featured in Appendix B may serve as a good guide for flow-pressure ranges of various rotary pumps. As you will see, such ranges reach 2000 to 3000 psi, above which reciprocating pumps take over. Large screw pumps reach 8000 to 9000 gpm, at lower pressures, but most of the rotary pumps are somewhere under 1000 to 1500 gpm or less.

## APPLICATION POINT #4 — VISCOSITY

Centrifugal pumps are not recommended above approximately 500 SSU. The viscous friction becomes high and the flow and pressure reduction is dramatic. For example, above 1000 to 2000 SSU, there is practically no flow though the pump. The HI[1] has a chart for flow, head, and efficiency correction with viscosity. This is another good reference material to keep in your engineering department.

On the other hand, rotary displacement pumps may not be your first choice at very low viscosities. They are seldom applied under 300 to 500 SSU. (Reciprocating pumps are the exception to this, due to their very tight clearances; they maintain flow, with low slip, even at the water-like pumpages.) There are also some exceptions to this rule. Gear pumps are sometimes made with pressure-loaded endplates to minimize lateral clearance to the thickness of the liquid film. However, their radial clearances (between gears and case) cannot be made too tight in order to avoid metal-to-metal contact, unless differential pressures are low. Screw pumps are available for low viscosity liquids: two-screw designs, supported externally, and driven by the timing gears, can have very tight radial clearances and maintain low slip even at reasonably high pressures. The shafts must be oversized in such cases, otherwise the long rotors' span between bearings would cause sagging deflections and potential contact between the rotors and the casing.

## APPLICATION POINT #5 — CHEMISTRY AND MATERIALS OF CONSTRUCTION

From the corrosion standpoint, applications can be classified as non-corrosive, mildly corrosive, and highly corrosive. Rotary pump designs are very simple and inexpensive for the low corrosive applications. Cast or ductile iron is often used for such applications as oil or fuel transfer, where the corrosion is almost nonexistent since the pumped oil has excellent lubricating properties which also helps to prevent corrosion. The iron construction allows cost reduction, and also helps from the reliability standpoint: iron has significant amounts of free carbon in its grain structure, which provide good additional self-lubricating characteristics. If occasional contact between the steel gears and iron case takes place, it does not cause failures. Nevertheless, this property of the iron should not be over-abused. If excessive pressures are applied, and the rotors are deflected onto casing, wear will take place, but it will be gradual and more or less predictable.

For the corrosive applications, stainless steel construction is often used, but the pump should not have mating pairs made from the austenitic stainless steel such as 316ss, particularly at low viscosity, which will gall easily (for gear pumps this would cause gears seizures, either onto themselves or to a casing). A 17–4ph steel is a better choice, but martensitic steels would be the best selection. Martensitic steels (such as 440C) have reasonable anti-galling resistance, as well as good corrosion resistance (although not as good as austenitics). A gear pump often has a 316ss casing with 440C or 17–4ph gears and carbon or silicon carbide wear plates, to prevent lateral seizure. The shafts generally are sized with enough stiffness to prevent deflection under maximum allowable design pressure onto casing, thus, ensuring no contact, at least in theory. In reality, however, occasional contact may occur, and designs should account for this.

## APPLICATION POINT #6 — ABRASIVENESS

Abrasive applications are difficult for rotary pumps, except for the single-screw type (progressing cavity). For very abrasive applications, a progressing cavity pump elastomer is an excellent choice, providing that there is sufficient floor space (these pumps are long), and moderate temperatures (under 300 to 350°F). Centrifugal pumps are also made for the abrasive applications, utilizing rubber linings (such as fly-ash removal applications at the power plants), or hard metal linings (such as hard iron, with Brinnell hardness to 660 Bhn).

Abrasion is exponential with speed, so slower running pumps would last longer. As a rule of thumb, for the abrasive applications, do not run rotary pumps much faster than 300 to 400 RPM. These slower speeds would require a gear reducer between the pump and a motor, unless a variable speed drive is used.

It may be useful to know the size of the abrasive particles — if the particles are too small, they will rub through the clearances, where wear is usually most detrimental. The large particles, however, would pass through the pump, and cause wear of the impellers (or gears, rotors, etc.) and casing, but this wear would be more predictable and not occur as quickly. In reality, however, there are doubtful applications, where the particles are only a certain size — usually, these large particles crush and break apart, and the pumpage carries along a wide variety of particles of all sizes. The bottom line is, if there is abrasive material present in the pumpage, it will probably do damage to the pump, sooner or later. Slow speeds, rubber linings, and harder materials will delay this process, to a tolerable degree.

## APPLICATION POINT #7 — TEMPERATURE

Under approximately 150°F, the application can be considered normal; between 150 to 250°F, it is moderate; and above 300°F, it is hot. For hot applications, elastomers are the determining factors, and appropriate selection charts are available. Use these charts with care — they usually contain *static* ratings, for "laboratory" conditions. Pump operating conditions are far more severe than laboratory conditions, and the temperature rating should be derated somewhat, perhaps by 50 to 70°F.

For very hot applications, 700 to 800°F, you may consider API-coded centrifugal pumps, or certain designs of the multiple-screw pumps. In any case, designs would contain opened-up clearances, external cooling methods (jackets), as well as requiring special start-up and shut-down procedures to ensure gradual warm-up and non-stratified fluid distribution inside the pump, and to ensure no sudden contact between the rotating and stationary parts, which could cause catastrophic seizure and failure.

## APPLICATION POINT #8 — SELF-PRIMING

Centrifugal pumps cannot "lift" the fluid, unless special self-priming designs are applied. Self-priming is not a forte of centrifugal pumps. Rotary pumps can prime very well, but designs require special features, such as tight clearance with wear plates and minimized cross-drilling, in order to avoid short-circuited air passage from discharge back to suction, and other problems. As a rule of thumb, three tiers of lift capabilities can be considered:

1 to 2 ft lift: Almost any rotary pump can provide this, as a standard, with no special design accommodations.

3 to 6 ft lift: Tighter clearance must be used, and the standard "off-the-shelf" rotary pump will probably not work. The cost of such special designs could be roughly twice the standard design.

6 to 15 ft lift: Very special approach to design, pressure loaded end-plates, tight radial clearances, oversized rotors to minimize deflection, etc. These pumps may cost 5 to 10 times more.

For lift values over 15 to 25 ft, the progressing cavity or diaphragm pumps should be considered.

## APPLICATION POINT #9 — DRIVER

Consider your power availability. For the electric motors, specify frequency, voltage, and power. Ensure the driver is sized for a complete operating range: remember from the previous chapters that the thixotrophic fluids may require more torque at the lower shear rate (i.e., near the start-up). Usually the AC motors are specified — they are more popular, reliable, and inexpensive. For very low powers, fractional power motors which are often DC, are used. Consider how pump flow will be varied (even if in the future): DC drives are easy to control, but the cost of the variable frequency drives (VFD) for the AC motors has been dropping steadily, and their reliability has increased dramatically. A VFD for pump under 5 hp could cost around $500 to $700. Above 5 hp, a rule of thumb (as of 1998) is $150 per hp (i.e., a 50 hp VFD should be around $7000 to $8000, and would probably be half of that value in three to four years).

## APPLICATION POINT #10 — FOOD APPLICATIONS

For obvious reasons standards and specifications govern pumps applications in the food industry. The Food and Drug Administration (FDA) specification does not detail any particular requirements for pump types or design parameters; it covers only allowable construction materials. Typically, these materials include stainless steel, certain plastics, and certain grades of carbon. Another specification called 3-A is produced by the U.S. Dairy Association. It covers many design issues specific to pumps: self-cleaning capabilities, absence of internal crevices where bacteria may spread, seals and seal chamber dimensions, etc. Lobe pumps are often used in these applications — lobes do not contact (in theory), and they are driven by the external timing gears. Absence of contact prevents any material shavings to pass on with the pumpage, thus contaminating the product. Shafts are robust, as is the rest of the pump, to ensure low deflections of rotors, resistance to piping loads, etc.

The above listed rules are general in nature, and each application has its nuances and specifics. It may be prudent for the plant personnel responsible for selection and specification of pumps to have a detailed checklist that would contain these and other points for the pump manufacturer to address during the quotation period. Some specifications such as API and PIP (the latter a recent specification within the process petrochemical industry — Process Industry Practices) already have similar checklists. It is advisable to have the latest versions of these and other relevant documents at your local facility.

# 15 Closing Remarks

The interaction between the pump and a system is a complex phenomenon. All factors, major and minor, must be addressed. The problem is that it is difficult to know, at the beginning, which of these factors are major and which are minor. It is always clear at the end what they are, but rarely at the beginning, when it is most needed to save time. Training and understanding the basic principles of what makes the pumps tick are prerequisites for successful troubleshooting. In the last several years, heightened attention to the equipment reliability and increase in MTBF has revived interest in better understanding and appreciation of pumping equipment. With a multitude of pump types operating in vastly different applications, successful plant operation is directly related to the attention given to the pumping equipment, which, as we learned in this book, can be tricky and stubborn, and must be approached systematically and diligently.

The author and publisher of this book hope the practical methods described here will help practicing engineers, plant operators, and maintenance personnel solve actual problems in their daily work. Understanding pump fundamentals should make their troubleshooting efforts easier and more rewarding. As a teaching reference, this book will be useful to college students in mechanical, chemical, and environmental disciplines. We believe that, for a technical and technologically-oriented environment, theory must continue to be tightly linked with practical and applied needs, creating an important and useful foundation for the engineering profession.

In writing this book, I tried to pay close attention to the accuracy, simplicity, and consistency of the material; however, it is difficult to avoid mistakes and controversy. I would be very grateful for any comments, criticisms, corrections, or suggestions that might aid in the creation or subsequent editions. Please send all remarks to me at the following addresses:

Dr. Lev Nelik, P.E.
140 Bedford Drive
Athens, GA 30606

nelik@compuserve.com

# Appendix A:
# Nomenclature

| | | |
|---|---|---|
| p | = | pressure (psi) |
| $p_d$ | = | pump discharge pressure (psi or psia) |
| $p_s$ | = | pump suction pressure (psig or psia) |
| $\Delta p$ | = | pump differential pressure (psi) |
| Q | = | pump flow (gpm) |
| q | = | pump unit flow (gal/rev or gpr) |
| FHP | = | fluid horsepower (hp) |
| BHP | = | break horsepower (hp) |
| $\eta$ | = | total pump efficiency |
| $\eta_H$ | = | hydraulic efficiency |
| H | = | pump head (general symbol) |
| Hi | = | pump ideal head (ft) |
| Ha | = | pump actual head (ft) |
| Hso | = | pump head at shut-off (shut valve) (ft) |
| $H_{sys}$ | = | system head (ft) |
| $h_{loss}$ | = | hydraulic losses (ft) |
| $V_S$ | = | velocity in suction (Vs), and discharge (Vd) pipe (ft/sec) |
| $\dfrac{V^2}{2g}$ | = | velocity head (ft) |
| g | = | gravitation constant (32.2 ft/sec²) |
| Zd | = | discharge side liquid level elevation above pump centerline (ft) |
| Zs | = | suction side liquid level elevation above pump centerline (ft) |
| NPSH | = | net positive suction head (ft) |
| NPSHA | = | available NPSH (ft) |
| NPSHR | = | NPSH required by the pump (ft) |
| NPSHR$_{3\%}$ | = | NPSH when 3% head drop had occurred (ft) |
| NPSHi | = | NPSH incipient, when first vapor bubbles begin to form (ft) |
| Ns | = | $\dfrac{N\sqrt{Q}}{H^{.75}}$ specific speed |
| Nss | = | $\dfrac{N\sqrt{Q}}{NPSHR^{.75}}$ suction specific speed |
| $\Omega_s$ | = | universal specific speed |
| OD | = | $D_2$ impeller outside diameter, (in) |
| $\alpha_f$ | = | absolute flow angle (degrees) |
| $\beta_f$ | = | relative flow angle (degrees) |
| $\beta_b$ | = | blade angle in the relative (W) direction (degrees) |
| $Ax_2$ | = | impeller exit area in relative direction (in²) |

| | | |
|---|---|---|
| Am | = | impeller meridional area (in²) |
| Afluid | = | 4eDPs: fluid area for the progressing cavity pump cross section |
| Pr | = | PC pump rotor pitch (in) |
| Ps | = | PC pump stator pitch (in |
| e | = | PC pump eccentricity (in) |
| $P_{s_{min}}$ | = | minimum required suction pressure for gear pumps (psia) |
| T | = | torque (in × lbs) |
| N | = | RPM = rotating speed |
| SSU | = | units Saybolt viscosity |
| cSt | = | viscosity in centistokes (approximately = SSU/5, for SSU >100) |
| ρ | = | fluid density (lbm/ft³) |
| γ | = | fluid specific weight (lbf/ft³) |
| $\gamma_o$ | = | specific weight for water at room temperature (lbf/ft³) |
| SG | = | $\gamma/\gamma_o$ specific gravity |
| y | = | shaft deflection (in) |
| L | = | bearing span (in) |
| V, $V_m$, $V_\theta$,  W, $W_\theta$, U | = | components of the velocity triangles for the centrifugal impeller |

# Appendix B: Conversion Formulas

This book emphasizes the fundamentals of different types of pumps, their similarities, and their differences. The number of formulas is not overwhelming, and the few formulas used are simple and straightforward. The user should have no difficulties understanding the formulas and their derivations. The units used in the book are U.S. system units. Listed below are a few formulas — covering major pump variables, such as flow, pressure, and power — for converting U.S. system units of measure to metric units:

*Flow*:

$$\frac{\text{GPM}}{4.403} = \text{m}^3/\text{HR}, \quad \frac{\text{GPM}}{15.9} = \text{liters/sec}, \quad \frac{\text{GPM(US)}}{1.2} = \text{GPM (Imp)}$$

*Pressure and Head*:

$$\frac{\text{psi}}{14.7} = \text{atmospheres}, \quad \frac{\text{psi}}{14.5} = \text{BARS}, \quad \frac{\text{psi}}{145} = \text{MPa}$$

*Power*:

$$\text{HP} \times 0.746 = \text{KW}$$

Other conversion formulas, if required, may be found in most engineering books and tables.[16]

# Appendix C:
# Rotary Pump Coverage Guide

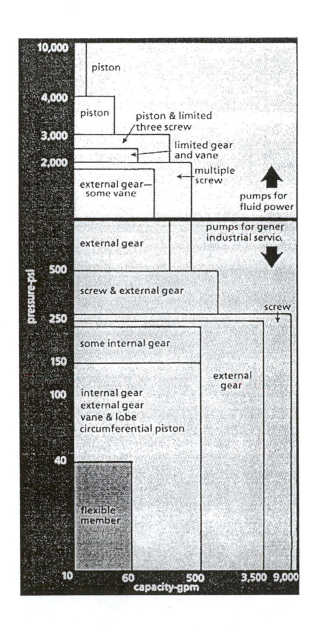

# References

1. Nelik, L., Pumps, *Kirk-Othmer Encyclopedia of Chemical Technology*, Vol. 4, 4th ed, John Wiley & Sons, New York, 1996.
2. Hydraulic Institute, *Hydraulic Institute Standards for Centrifugal, Rotary & Reciprocating Pumps,* Parsippany, NJ, 1994.
3. Russell, G. *Hydraulics*, 5th ed., Henry Holt and Company, New York, 1942.
4. Stepanoff, A. J. *Centrifugal and Rotary Pumps*, 2nd ed., John Wiley & Sons, New York, 1948.
5. Shigley, J. and Mischke, C., Gears, *Mechanical Engineering Design*, 5th ed., McGraw-Hill, New York, 1989.
6. Avallone, E., Hydrodynamics Bearings, in *Marks' Handbook for Mechanical Engineers, Section 8*, 9th ed., McGraw-Hill, New York, 1986.
7. Budris, A., Preventing Cavitation in Rotary Gear Pumps, *Chemical Engineering*, May 5, 1980.
8. SKF, *SKF General Catalog No. 4000US (Bearings),* King of Prussia, PA, 1991.
9. Luer, K. and Marder, A., Wear Resistant Materials for Boiler Feed Pump Internal Seals, *Advances in Steam Turbine Technology for Power Generation,* ASME Reprint from PWR Book No. G00518, Vol. 10, 1990.
10. Nelik, L., Positive Displacement Pumps, paper presented at the Texas A&M 15th Int. Pump Users Symp., section on Screw Pumps by J. Brennan, Houston, March, 1998.
11. Dillon, M. and Vullings, K., Applying the NPSHR Standard to Progressing Cavity Pumps, *Pumps and Systems*, 1995.
12. Bourke, J., Pumping Abrasive Liquids with Progressing Cavity Pumps, J. Paint Tech., Vol. 46, Federation of Societies for Paint Technologies, Philadelphia, PA, August, 1974.
13. Platt, R., Pump Selection: Progressing Cavity, *Pumps and Systems*, August, 1995.
14. Nelik, L., Progressing Cavity Pumps, Downhole Pumps, and Mudmotors: Geometry and Fundamentals (in press).
15. Schlichting, H., *Boundary-Layer Theory*, 7th ed., McGraw-Hill, New York, 1979.
16. Heald, C., *Cameron Hydraulic Data,* 18th ed., Ingersoll-Dresser Pumps, Liberty Corner, NJ, 1996.
17. *The Sealing Technology Guidebook*, 9th ed., Durametallic Corp., Kalamazoo, MI, 1991.
18. Specification for Horizontal End Suction Centrifugal Pumps for Chemical Process, ANSI/ASME B73.1M-1991 Standard, ASME, New York, 1991.
19. Block, H., *Process Plant Machinery*, Butterworth & Co. Publishers Ltd., Kent, U.K., 1989.
20. Karassik, I., et al., *Pumps Handbook*, McGraw-Hill, New York, 1976.
21. Cholet, H., *The Progressing Cavity Pumps*, Editions Technip, France, 1996.

22. Fritsch, H., *Metering Pumps: Principles, Designs, Applications*, 2nd ed., Verlag Moderne Industrie, Germany, 1994.

23. Florjancic, D., Net Positive Suction Head for Feed Pumps, *Sulzer Report*, 1984.

24. Feedpump Operation and Design Guidelines, Summary Report TR-102102, Sulzer Brothers and EPRI, Winterthur, Switzerland, 1993.

25. Nelik, L., How Much NPSHA is Enough?, *Pumps and Systems*, March, 1995.

26. Varga, J., Sebestyen, G., and Fay, A., Detection of Cavitation by Acoustic and Vibration Measurement Methods, La Houville Blancha, 1969.

27. Kale, R. and Sreedhar, B., A Theoretical Relationship Between NPSH and Erosion Rate for a Centrifugal Pump, Vol. 190, ASME FED, 1994, 243.

28. Nelik, L., Salvaggio, J., Joseph, J., and Freeman, J., Cooling Water Pump Case Study — Cavitation Performance Improvement, paper presented at the Texas A&M Int. Pump Users Symp., Houston, TX, March, 1995.

29. API Standard 610, Centrifugal Pumps for General Refinery Service, 8th ed, Washington, D.C., 1995.

30. Frazer, H., Flow Recirculation in Centrifugal Pumps, presented at the ASME Meeting, 1981.

31. Karassik, I., Flow Recirculation in Centrifugal Pumps: From Theory to Practice, presented at the ASME Meeting, 1981.

32. The Characteristics of 78 Related Airfoil Sections from Tests in the Variable Density Wind Tunnel, Report No. 460, NACA.

33. Florjancic, D., Influence of Gas and Air Admissions the Behavior of Single- and Multi-Stage Pumps, *Sulzer Research*, No. 1970.

34. Nelik, L. and Cooper, P., Performance of Multi-Stage Radial-Inflow Hydraulic Power Recovery Turbines, ASME, 84-WA/FM-4.

# Index